Methods in Molecular Biology

Series Editor
John M. Walker
School of Life and Medical Sciences
University of Hertfordshire
Hatfield, Hertfordshire, AL10 9AB, UK

For further volumes:
http://www.springer.com/series/7651

The Human Virome

Methods and Protocols

Edited by

Andrés Moya

Institute for Integrative Systems Biology, University of València, València, Spain; Foundation for the Promotion of Health and Biomedical Research of the Valencian Community (FISABIO), València, Spain; CIBER in Epidemiology and Public Health (CIBEResp), Madrid, Spain

Vicente Pérez Brocal

Foundation for the Promotion of Health and Biomedical Research of the Valencian Community (FISABIO), València, Spain; CIBER in Epidemiology and Public Health (CIBEResp), Madrid, Spain

Humana Press

Editors
Andrés Moya
Institute for Integrative Systems Biology
University of València
València, Spain

Foundation for the Promotion of Health
and Biomedical Research of the Valencian
Community (FISABIO)
València, Spain

CIBER in Epidemiology and Public Health
(CIBEResp)
Madrid, Spain

Vicente Pérez Brocal
Foundation for the Promotion of Health
and Biomedical Research of the Valencian
Community (FISABIO)
València, Spain

CIBER in Epidemiology and Public
Health (CIBEResp)
Madrid, Spain

ISSN 1064-3745 ISSN 1940-6029 (electronic)
Methods in Molecular Biology
ISBN 978-1-4939-8681-1 ISBN 978-1-4939-8682-8 (eBook)
https://doi.org/10.1007/978-1-4939-8682-8

Library of Congress Control Number: 2018948848

Printed on acid-free paper

This Humana Press imprint is published by the registered company Springer Science+Business Media, LLC part of Springer Nature.
The registered company address is: 233 Spring Street, New York, NY 10013, U.S.A.

Preface

The virome or viral metagenome consists of the genomes of all the viruses that inhabit a particular organism or environment. In particular, the human virome constitutes the viral component of the human microbiome and has been defined as "the collection of all viruses found in or on humans, including viruses causing acute, persistent or latent infection, and viruses integrated into the human genome. It includes both eukaryotic and prokaryotic viruses (bacteriophages)" [1].

The relevance of viruses is patently obvious as they exert direct or indirect impact on human health and disease [1, 2]. For instance, eukaryotic viruses may provoke either acute or persistent infections, as well as interact with the immune system and other pathogens. Also, prokaryotic viruses may show certain associations with health and disease, even though less direct, by their effects on the bacterial communities' structure and metabolic potential, as well as by carrying pathogenic potential toxins or antibiotic resistance genes.

Notwithstanding the ubiquity and relevance of the ecological role that viruses play in the human body, their study from a metagenomic perspective has taken off only recently and, consequently, has entailed a relative delay compared to studies focused on bacterial counterparts, most likely because of technical limitations that the isolation and characterization of viruses encompass [3]. Those include, but are not limited to, the incompleteness of viral genomic databases, the mosaicism of viral genomes due to recurrent horizontal gene transfer events, the high sequence variation and evolutionary rates, due to accumulation of mutations, and the lack of universally conserved genes (unlike the 16S ribosomal RNA gene, used as universal marker for prokaryotes).

Traditional strategies to characterize viruses based on culture, amplification by polymerase chain reaction (PCR), or microscopy/hybridization imply a series of limitations, such as being restricted to a particular virus or group of viruses for which known information is previously available; or are techniques that rely on homology to known viruses, when the reality is much more complex. Finally, as with prokaryotes and eukaryotes, many viruses remain uncultivable, and thus they can only be detected by culture-independent approaches. Metagenomic strategies can cope with some of these weaknesses, although they are not free of limitations that challenge these approaches [4].

Therefore, sample preparation stands for a crucial step in downstream analyses, including next-generation sequencing and computational analyses meant to deal with the disentanglement of the hodgepodge of raw data generated through these novel approaches.

The aim of this book is to compile the most comprehensive and up-to-date collection of cutting-edge protocols used in wet and dry lab to investigate the viral communities harbored within and on the human body. A practical and experience-based book, it includes methods for collection, isolation, identification, and computational/statistical analysis of the human virome. This volume is arranged by body niches to cover those methodological issues inherent to the human tissues and organs, facilitating the researchers their daily work from beginning to end in the field of the research on the human virome.

València, Spain — *Andrés Moya*
València, Spain — *Vicente Pérez-Brocal*

References

1. Wylie KM, Weinstock GM, Storch GA (2012) Emerging View of the Human Virome. Trans Res 160:283–290
2. De Paepe M, Leclerc M, Tinsley CR, Petit MA (2014) Bacteriophages: an underestimated role in human and animal health? Front Cell Infect Microbiol 4:39
3. Ogilvie LA, Jones BV (2015) The human gut virome: a multifaceted majority. Front Microbiol 6:918
4. Beerenwinkel N, Günthard HF, Roth V, Metzner KJ (2012) Challenges and opportunities in estimating viral genetic diversity from next-generation sequencing data. Front. Microbiol 3:1–16

Contents

Contributors

JEREMY J. BARR • *School of Biological Sciences, Monash University, Clayton, VIC, Australia*
NATASHA BONILLA • *Department of Biology, San Diego State University, San Diego, CA, USA*
JUAN VICENTE BOU • *Department of Genetics, Institute for Integrative Systems Biology, Universitat de València, Paterna, Spain*
JOSUÉ L. CASTRO-MEJÍA • *Department of Food Science, University of Copenhagen, Copenhagen, Denmark*
ONDŘEJ CINEK • *Department of Pediatrics, Second Faculty of Medicine, Charles University, Prague, Czech Republic; Department of Medical Microbiology, Second Faculty of Medicine, Charles University, Prague, Czech Republic*
NÁDIA CONCEIÇÃO-NETO • *Laboratory of Viral Metagenomics, Department of Microbiology and Immunology, Rega Institute, KU Leuven – University of Leuven, Leuven, Belgium; Laboratory of Clinical Virology, Department of Microbiology and Immunology, Rega Institute, KU Leuven – University of Leuven, Leuven, Belgium*
LING DENG • *Department of Food Science, University of Copenhagen, Copenhagen, Denmark*
CHRISTELLE DESNUES • *Unité de Recherche sur les Maladies Infectieuses Tropicales Emergentes (URMITE), IHU Méditerranée Infection, Assistance-Publique des Hôpitaux de Marseille, Aix-Marseille Université, CNRS 7278, IRD 198, INSERM1095, Marseille, France*
PILAR DOMINGO-CALAP • *Department of Genetics, Institute for Integrative Systems Biology, Universitat de València, Paterna, Spain*
MÁRIA DŽUNKOVÁ • *Australian Centre for Ecogenomics, University of Queensland, St. Lucia, QLD, Australia*
XAVIER FERNANDEZ-CASSI • *Department of Genetics, Microbiology and Statistics, Faculty of Biology, University of Barcelona, Barcelona, Spain*
RODRIGO GARCÍA-LÓPEZ • *Department of Genomics and Health, Fundación para el Fomento de la Investigación Sanitaria y Biomédica de la Comunitat Valenciana (Fisabio), Valencia, Spain*
ALLAN R. GLANVILLE • *The Lung Transplant Unit, St. Vincent's Hospital, Sydney, NSW, Australia*
STEPHEN M. GROSS • *Illumina, Inc., San Diego, CA, USA*
JUN HANG • *Viral Diseases Branch, Walter Reed Army Institute of Research, Silver Spring, MD, USA*
RICHARD G. JARMAN • *Viral Diseases Branch, Walter Reed Army Institute of Research, Silver Spring, MD, USA*
BRYN Y. KIMURA • *Computational Biology, Faculty of Medicine and Life Sciences, University of Tampere, Tampere, Finland*
CLAUDIA KOHL • *Centre for Biological Threats and Special Pathogens, Robert Koch Institute, Berlin, Germany*

LENKA KRAMNÁ • *Department of Pediatrics, Second Faculty of Medicine, Charles University, Prague, Czech Republic; Department of Medical Microbiology, Second Faculty of Medicine, Charles University, Prague, Czech Republic*

ANDREAS KURTH • *Centre for Biological Threats and Special Pathogens, Robert Koch Institute, Berlin, Germany*

JAKE LIN • *Computational Biology, Faculty of Medicine and Life Sciences, University of Tampere, Tampere, Finland; Virology, Faculty of Medicine and Life Sciences, University of Tampere, Tampere, Finland*

ALBERTO LÓPEZ-BUENO • *Centro de Biología Molecular Severo Ochoa (Universidad Autónoma de Madrid-Consejo Superior de Investigaciones Científicas), Madrid, Spain*

JOSE MANUEL MARTÍ • *Institute for Integrative Systems Biology (I2SysBio), Parc Científic de la Universitat de València, Valencia, Spain*

SANDRA MARTÍNEZ-PUCHOL • *Department of Genetics, Microbiology and Statistics, Faculty of Biology, University of Barcelona, Barcelona, Spain*

JELLE MATTHIJNSSENS • *Laboratory of Viral Metagenomics, Department of Microbiology and Immunology, Rega Institute, KU Leuven – University of Leuven, Leuven, Belgium*

DAVID MAYO-MUÑOZ • *Danish Archaea Centre, Department of Biology, University of Copenhagen, Copenhagen, Capital Region, Denmark*

ALICIA B. MITCHELL • *The Woolcock Institute of Medical Research, Sydney, NSW, Australia; University of Technology Sydney, Sydney, NSW, Australia; The Lung Transplant Unit, St. Vincent's Hospital, Sydney, NSW, Australia*

SONIA MONTEIL-BOUCHARD • *Unité de Recherche sur les Maladies Infectieuses Tropicales Emergentes (URMITE), IHU Méditerranée Infection, Assistance-Publique des Hôpitaux de Marseille, Aix-Marseille Université, CNRS 7278, IRD 198, INSERM1095, Marseille, France*

DENNIS S. NIELSEN • *Department of Food Science, University of Copenhagen, Copenhagen, Denmark*

MATTI NYKTER • *Computational Biology, Faculty of Medicine and Life Sciences, University of Tampere, Tampere, Finland*

SAMI OIKARINEN • *Computational Biology, Faculty of Medicine and Life Sciences, University of Tampere, Tampere, Finland; Virology, Faculty of Medicine and Life Sciences, University of Tampere, Tampere, Finland*

MARCOS PARRAS-MOLTÓ • *Centro de Biología Molecular Severo Ochoa (Universidad Autónoma de Madrid-Consejo Superior de Investigaciones Científicas), Madrid, Spain*

ALEJANDRO REYES • *Department of Biological Sciences, Universidad de los Andes, Bogotá, Colombia; Center for Genome Sciences and Systems Biology, Washington University School of Medicine, St. Louis, MO, USA*

MARTA RUSIÑOL • *Department of Genetics, Microbiology and Statistics, Faculty of Biology, University of Barcelona, Barcelona, Spain*

TASHA M. SANTIAGO-RODRIGUEZ • *ATCC-Center for Translational Microbiology, Institute for Life Science Entrepreneurship, Union, NJ, USA*

GARY P. SCHROTH • *Illumina, Inc., San Diego, CA, USA*

SARAH TEMMAM • *Biology of Infection Unit, Laboratory of Pathogen Discovery, Institut Pasteur, INSERM U1117, Paris, France*

MARC VAN RANST • *Laboratory of Clinical Virology, Department of Microbiology and Immunology, Rega Institute, KU Leuven – University of Leuven, Leuven, Belgium*

JORGE FRANCISCO VÁZQUEZ-CASTELLANOS • *Department of Genomics and Health, Fundación para el Fomento de la Investigación Sanitaria y Biomédica de la Comunitat Valenciana (Fisabio), Valencia, Spain*

FINN K. VOGENSEN • *Department of Food Science, University of Copenhagen, Copenhagen, Denmark*

SHANNON D. WALLS • *Viral Diseases Branch, Walter Reed Army Institute of Research, Silver Spring, MD, USA*

YU YANG • *Viral Diseases Branch, Walter Reed Army Institute of Research, Silver Spring, MD, USA*

KWE CLAUDE YINDA • *Laboratory of Viral Metagenomics, Department of Microbiology and Immunology, Rega Institute, KU Leuven – University of Leuven, Leuven, Belgium; Laboratory of Clinical Virology, Department of Microbiology and Immunology, Rega Institute, KU Leuven – University of Leuven, Leuven, Belgium*

Part I

General Experimental Methods

Chapter 1

Flow Cytometry and Direct Sequencing of Viruses

Mária Džunková

Abstract

In order to describe a novel uncultured viral species, it is essential to obtain the DNA sequence of their full genomes. The ability to distinguish the viral genome from the genome of its bacterial host is the major challenge of the modern viromics. The major obstacles for mining of viral genomes in metagenomic assemblies is bacterial contamination in viromes and low DNA input for sequencing.

These obstacles can be overcome by flow cytometry that allows collecting free viral particles from environmental samples. In addition, fluorescence activated cell sorting reduces the bacterial contamination. By using optimized sequencing protocols, the ultra-low input DNA samples can be sequenced directly, without the need for whole genome amplification. This chapter provides details for staining of environmental viruses, flow cytometry, and direct sequencing of ultra-low input DNA samples on Illumina platform.

Key words Flow cytometry, Ultra-low input DNA samples, Viruses, Whole genome amplification, DNA sequencing

1 Introduction

Bacteriophages are the most abundant biological entities. Mining of large metagenomic assemblies performed in recent years allowed discoveries of thousands of novel unculturable environmental viruses, mostly bacteriophages [1–5].

However, the environmental samples purified by filtration or by density gradients can still contain some contaminating bacteria, thus the putative viral contigs may actually come from bacterial cells [6]. This is unacceptable for many research projects which require direct experimental evidence that novel uncultured viruses can exist as free infectious particles not integrated into bacterial genomes. That is the point when the flow cytometry (FC) analysis comes into play [7].

FC is a high-throughput screening technique allowing us to determine size and fluorescence levels of cells or viral particles resuspended in a buffer. The most common method for

Andrés Moya and Vicente Pérez Brocal (eds.), *The Human Virome: Methods and Protocols*, Methods in Molecular Biology, vol. 1838, https://doi.org/10.1007/978-1-4939-8682-8_1, © Springer Science+Business Media, LLC, part of Springer Nature 2018

distinguishing between bacteria and viruses is based on the notion that the average size of a bacterium is above 0.5 µm and the size of an average virus is below 0.1 µm [8]. The parameters and sorting gates for the fluorescence activated cell sorting (FACS) can be adjusted according to the sample specificities, e.g., some samples can contain bacteria of size below 0.2 µm [9], while other samples can contain giant viruses of size over 0.7 µm [10]. FACS allows us to set several sorting gates corresponding to viruses of different sizes and so to study diversity of subsets of a virome. The biggest advantage of the flow cytometry is its ability to separate single viral particles [11, 12].

The number of viral particles in subsetting gates collected by FACS is often so low that the amount of extracted DNA is below detection limit of the commonly used PicoGreen assay (<0.25 pg/µL). It means that such samples do not contain the minimal DNA amount required for the preparation of a sequencing library. DNA coming from single viral particles separated by FACS can be enriched by whole genome amplification (WGA) methods [13]. However, despite recent improvements in the WGA methods, the resulting DNA will still consist of overamplified regions, while other regions may not be amplified at all [14]. While quality of a single amplified viral genome can be assessed relatively easily [12], it may be difficult in case of amplified bulks containing several hundreds of viral particles sorted from different FACS gates.

Nevertheless, it was demonstrated that the usage of WGA is not needed for bulk samples containing hundreds of cells (or viral particles). If the concentration of extracted DNA does not reach the minimal limit required by the sequencing library preparation protocols, the sample can still be successfully sequenced on the selected sequencing platform, if the concentration of the completed library contains the number of molecules required for loading on the sequencing plate [15–17]. It was demonstrated that using an optimized sequencing library preparation protocol, a sample containing a few hundreds of viral particles can be sequenced directly, without any whole genome amplification [18].

The protocol in this chapter shows details for purification of viral particles from human faecal samples, their fluorescent staining, setting of FC parameters for FACS sorting, DNA extraction and Illumina sequencing library preparation.

The protocol provides basis for setting up the FACS visualization thresholds and sorting gates; however, these parameters must be set up individually for each sorting equipment. The user will also select the sorting gates according to the research project objective (e.g., several subsets of a virome can be sorted at once to assess diversity of viruses of different particle size and genome size). In our experience, the sorting gates should contain at least 100 viruses in order to obtain a successful sequencing run without using the whole genome amplification.

The sorting of the cells may be followed by DNA extraction described in this chapter adapted from Thurber et al. [19]; however, some specific viruses may require a specialized DNA extraction protocol. Independently of the DNA extraction protocol used, the extracted DNA must be finally resuspended in the smallest volume possible (not more than 20 μL), as it will be followed to the optimized sequencing library preparation protocol. The protocol for an optimized Nextera library preparation (Nextera XT DNA Sample Preparation Kit, Document # 15031942 v02 [20]) for ultra-low input samples has been adapted from Rinke et al. [17].

2 Materials

2.1 Preparation of Viruses for FACS

1. *E. coli* strain ER2738.
2. M13KE phage.
3. LB medium: Resuspend 10 g tryptone, 5 g yeast extract and 5 g NaCl in 1 L water and autoclave.
4. Salt solution (SS): Resuspend 9 g of NaCl in 1 L of water and autoclave.
5. Sterile water (autoclaved, filtered through 0.1 μm filters). Distribute the water into 1.5 mL tubes and UV-radiate them for 60 min.
6. 50 mL tubes (at least 27 pcs).
7. Scalpel blade.
8. Chloroform.
9. Serological pipettes, at least 7 pcs.
10. 1.5 mL tubes, at least 14 pcs.
11. 2 mL tubes, at least 120 pcs.
12. 0.22 μm pore filters, at least 12 pcs.
13. 0.1 μm pore filters, at least 6 pcs.
14. 10 mL syringes, at least 18 pcs.
15. Polyethylene glycol (PEG): Resuspend 200 g of PEG-6000 and 150 g NaCl in 1 L of hot water and autoclave. If the layers separate, mix them well and cool down in the fridge. Prepare a stock for the viral precipitation—filtrate through 0.22 μm pore filters.
16. Syto 9 nucleic acids stain.
17. Eppendorf Low-bind tubes (1.5 mL) for collecting viral particles by FACS.

2.2 DNA Extraction of Viruses

1. DNase.
2. Lysozyme. Prepare it at conc. 10 mg/mL in sterile water.
3. SDS 10%. Resuspend 10 g of sodium dodecyl sulphate (SDS) in 100 mL of sterile water. UV-radiate for 60 min.
4. Proteinase K, PCR grade.
5. CTAB: Add 1 g of NaCl to 15 mL of water and slowly add 2.5 g of cetyltrimethylammonium bromide (CTAB). Heat if necessary for dissolving. Filter through 0.1 μm pore filters and UV-radiate.
6. Phenol.
7. Phenol:Chloroform:Isoamyl Alcohol 25:24:1 saturated with 10 mM Tris–HCl, pH 8.0, 1 mM EDTA.
8. 5 M NH_4-acetate: Resuspend 385.4 g of NH_4-acetate in 1 L of water. Autoclave, prepare a small stock, filter through 0.1 μm pore filters, and UV-radiate.
9. 70% ethanol, filter through 0.1 μm pore filters and UV-radiate.
10. 100% ethanol, filter through 0.1 μm pore filters and UV-radiate.
11. Two additional low bind tubes of 1.5 mL.

2.3 Optimized Sequencing Library Preparation Protocol for Ultra-Low Input DNA Samples

1. Nextera XT DNA Sample Preparation Kit (Illumina), or other kit from this series according to the user's sequencing needs.
2. Magnetic stand, e.g., Agencourt SPRIPlate Super Magnet Plate.
3. Ampure XP beads.
4. 80% ethanol.
5. KAPA Library Quantification Kits (or other kit from this series according to the user's qPCR equipment in the laboratory).

3 Methods

3.1 Preparing Negative Controls

1. Prepare three falcon tubes containing 30 mL of LB medium. The tubes will contain following overnight cultures incubated at 37 °C using medium shaking speed:
   - First tube inoculated with 1 μL of *E. coli* strain ER2738 and 0.5 μL M13KE phage.
   - Second tube inoculated with 1 μL of *E. coli* strain ER2738 without any phage.
   - Third tube containing only 30 mL LB.
2. Prepare a negative faecal sample negative control: 50 mL tube containing 30 mL of SS.

3.2 Preparation of the Faecal Samples

1. Work in a biosafety hood. Cut two pieces of faecal sample of approx. volume 1 mL by a scalpel blade and put them into two 50 mL tube containing 30 mL of SS and vortex rigorously.
2. Centrifuge the two tubes at 200 × *g* for 2 min at 4 °C and transfer the supernatants into new 50 mL tubes.

3.3 Processing of the Viruses

1. Centrifuge the negative control tubes (4 tubes) and faecal sample tubes (2 tubes) at 3,270 × *g* for 15 min at 4 °C.
2. Transfer the supernatants into new 50 mL tubes.
3. Collect about 10 μL of pellet of one the faecal samples and also of the *E. coli* not infected by phage and infected by phage. Store these pellets in 1.5 mL tubes in the fridge, as they will serve as bacterial controls in the flow cytometry analysis.
4. Take one tube containing the faecal sample and add 300 μL of chloroform. Vortex the tube for 1 min and centrifuge for 10 min at 3,270 × *g* at 4 °C. Transfer the supernatant by a serological pipette into a new tube (*see* **Note 1**).
5. Centrifuge all the six tubes again for 10 min at 3,270 × *g* at 4 °C.
6. Collect the supernatant using serological pipettes. The supernatant from one 50 mL tube may fit into about ten 2 mL tubes.
7. Centrifuge the tubes at 27,000 × *g* for 15 min at 4 °C.
8. Collect approximately 1.7 mL of the supernatant from each tube and transfer it into a new 2 mL tube.
9. Centrifuge again all the tubes at 27,000 × *g* for 15 min at 4 °C (*see* **Note 2**).
10. Transfer the supernatants from each set of tubes (about 10 mL in total per each set) directly into 10 mL syringes with 0.22 μm filters attached on them.
11. Filter the supernatants directly into new 10 mL syringes attached to 0.22 μm filters. The filtrates from the second round of filtration should go directly into the last set of syringes with 0.1 μm filters attached.
12. The supernatant after the filtration through 0.1 μm pores filters should go directly into 50 mL tubes containing 2.5 mL of ice-cold PEG.
13. Incubate the sample with PEG overnight on ice.
14. After incubation, centrifuge the tubes for 1 h at 3,270 × *g* at 4 °C.
15. Discard the supernatant carefully. Resuspend the pellets (small almost transparent pellets) in 1 mL of filtered and UV sterilized water by pipetting up and down (*see* **Note 3**).

16. Transfer the resuspended pellets into 1.5 mL sterilized tubes.
17. Add 250 μL of filtered ice-cold PEG and vortex vigorously. Incubate on ice for 30 min (*see* **Notes 4** and **5**).
18. Centrifuged the tubes at 27,000 × *g* for 15 min at 4 °C. Discard the supernatants and resuspend the pellets by pipetting up and down in 1 mL of water.

3.4 Staining of the Samples

1. Split each of the six samples into halves to get stained and non-stained counterparts. Mark the counterparts as "neg" and "pos." Bring the volume of the 12 samples back to 1 mL with sterile water.
2. Use the bacterial pellets stored in the fridge from the step 3.3.3. Resuspend the bacterial pellet in 1 mL SS to get the optical density OD600 = 0.1 (*see* **Note 6**).
3. Add 1 μL of SYTO 9 stain to the tubes marked as "pos" and vortex (*see* **Note 7**).
4. Incubate all the tubes (both "pos" and "neg") in the dark at room temperature for 30 min.
5. Add 250 μL of PEG into all tubes. Vortex the sample and incubate for 30 min on ice (*see* **Note 8**).
6. Centrifuge the tubes for 15 min at 13,000 × *g* at 4 °C and discard the supernatants.
7. Resuspend the pellets in 1 mL of sterile water by pipetting up and down.

3.5 Flow Cytometry of the Viruses

1. Set a high visualization threshold to green fluorescence and the minimal threshold for the cell size (minimal threshold for forward scatter FSC or side scatter SSC). The recommended threshold setting for example for the Accuri flow cytometer (BD Biosciences) is FL1 2000 and FSC 200 (*see* **Note 9**).
2. Start processing the negative M13KE phage during 30 s on slow recording. After that, process the stained M13KE phage. Adjust the fluorescence threshold if necessary (*see* **Note 10**).
3. Flush the flow cytometer several times with sterile water and bleach. Continue only if the sterile water sample does not contain any fluorescent events.
4. Process first the negative samples in the following order:
   - SS.
   - LB.
   - Filtrate of *E. coli* not infected by phage.
   - Filtrate of *E. coli* infected by phage.
   - Faecal viral sample processed with chloroform.
   - Faecal viral sample processed without chloroform.

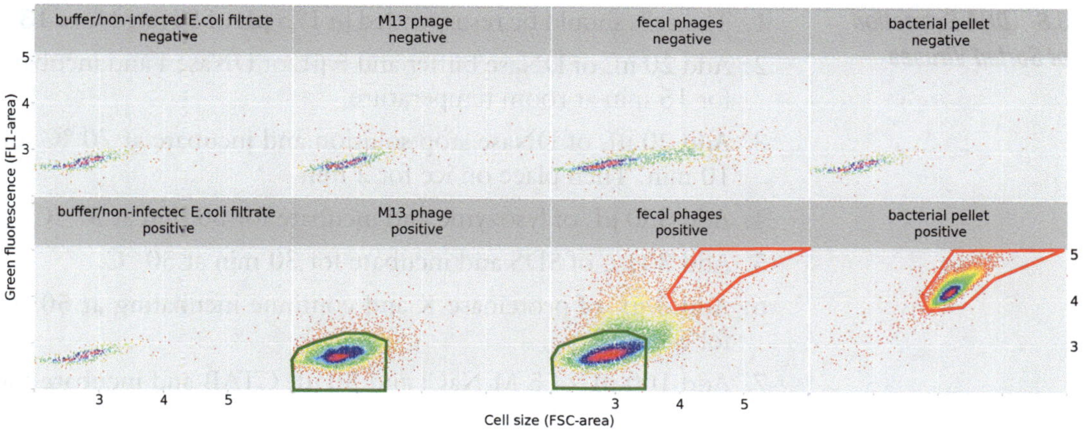

Fig. 1 An example of the flow cytometry plots of the viruses and negative controls. Flow cytometry bi-plots obtained on the Accuri flow cytometer. A high threshold has been set to fluorescence (FL1 = 2000), while the minimal threshold has been set to the size threshold (FSC = 200). The figures show that all the negative samples and the buffer/*E. coli* filtrate not infected by phages do contain only few events that pass the fluorescence threshold, thus they can be considered negative. These few events are actually so called "electrical noise" detected by the flow cytometer above the fluorescence threshold. The real viral particles appear in the bottom left part of the plots, while the bacterial cells appear in the upper right part of the plots, because the viral particles are smaller and contain less DNA than the bacterial cells. It is possible to see that the plot corresponding to positive (stained) faecal viruses contain particles of different sizes and fluorescence than the plot corresponding to the stained M13KE phage which is more uniform in size and fluorescence. The final sorting FACS gates may be set depending on the user's project objective, for example several different gates corresponding to viruses of different sizes may be sorted at once. The sorting gates for the faecal viruses should not reach the area corresponding to the bacterial cells (red gate) and should be preferentially within the area corresponding to the phage control (green gate)

- *E. coli* cells (pellet) not infected by phage.
- *E. coli* cells (pellet) infected by phage.
- Faecal bacterial cells (pellet).

5. Flush the flow cytometer several times with sterile water and bleach. Continue only if the sterile water sample does not contain almost any fluorescent events (*see* **Note 11**).
6. Afterward, process the positive (stained) samples in the same order as above. The resulting image should look similar as shown in Fig. 1.
7. Set the sorting gate of faecal viruses by comparison of the size and fluorescence of the bacterial cells and in comparison with the control phage M13. The faecal viruses should be of the same fluorescence and size as the control phage M13, while the sorting gate marking the viruses should not reach the area of the faecal bacterial cells (*see* **Note 12**).
8. Sort the faecal viruses into a sterile 1.5 mL tube. Set single-cell purity mode for the cell sorting, if possible.

3.6 DNA Extraction of Sorted Viruses

1. The cells should be resuspended in 175 μL of SS (*see* **Note 13**).
2. Add 20 μL of DNase buffer and 5 μL of DNase I and incubate for 15 min at room temperature.
3. Add 20 μL of DNase stop solution and incubate at 70 °C for 10 min. Then place on ice for 2 min.
4. Add 350 μL of lysozyme and incubate for 30 min at 37 °C.
5. Add 30 μL of SDS and incubate for 30 min at 50 °C.
6. Add 3 μL of proteinase K and continue incubating at 50 °C for 30 min.
7. Add 100 μL of 5 M NaCl and 80 μL CTAB and incubate for 15 min at 65 °C.
8. Add 700 μL of phenol and shake vigorously or vortex for 2 min. Centrifuge for 5 min at maximum speed.
9. Transfer the supernatant to the new tube containing 700 μL of phenol:chloroform and shake vigorously or vortex for 2 min. Centrifuge 5 min at maximum speed.
10. Transfer the supernatant to the new tube containing 220 μL of 5 M NH4-acetate and 2 μL of glycogen. Mix by inverting the tube up and down.
11. Add 500 μL of isopropanol and vortex to see bubbles. Incubate the sample at room temperature for 10 min and then store it at −80 °C for 2–18 h.
12. Centrifuge for 45 min at 4 °C at maximum speed.
13. Remove the supernatant. Add 500 μL of 70% ethanol. Vortex and centrifuge at maximum speed for 45 min at 4 °C (350 μL of 100% ethanol).
14. Discard the supernatant and centrifuge shortly to collect and discard all remaining ethanol.
15. Let remaining ethanol dry out for 30 min at room temperature at laminar flow hood.
16. Resuspend in 10 μL of water.

3.7 Preparation of the Nextera Sequencing Library

1. Dilute the Amplicon Tagment Mix (ATM) 10 times with sterile water (*see* **Note 14**).
2. Mix 10 μL of Tagment DNA Buffer (TD) with 5 μL of input DNA.
3. Add 50 μL of the diluted ATM. Pipette to mix.
4. Centrifuge at 280 × *g* at 20 °C for 1 min.
5. Place on the pre-programmed thermal cycler and run the following program:

   55 °C for 5 min, followed by hold at 10 °C, preheated lid option.

6. When the sample reaches 10 °C, immediately add 5 μL of the Neutralize Tagment Buffer (NT). Pipette to mix.
7. Centrifuge at 280 × *g* at 20 °C for 1 min.
8. Incubate at room temperature for 5 min.
9. Select the correct combination of Index 1 (i7) and Index 2 (i5) according to your choice and manufacturer's instructions (*see* **Note 15**).
10. Add 5 μL of Index 1 and then add 5 μL of Index 2 to the sample. Add 15 μL of Nextera PCR Master Mix (NPM). Pipette to mix.
11. Centrifuge at 280 × *g* at 20 °C for 1 min.
12. Place on the pre-programmed thermal cycler and run the following program with preheated lid option: 72 °C for 3 min, incubation at 95 °C for 30 s, followed by 20 cycles at 95 °C for 10 s, 55 °C for 30 s and 72 °C for 30 s, and final extension at 72 °C for 4 min. Hold at 10 °C (*see* **Note 16**).
13. Add 1.66× of AMPure XP beads (30 μL) and vortex.
14. Incubate at room temperature for 5 min, then place on a magnetic stand and wait until the liquid is clear (about 2 min).
15. Remove and discard the supernatant.
16. Wash two times as follows: Add 200 μL of fresh 80% ethanol. Incubate on the magnetic stand for 30 s. Remove and discard all the supernatant.
17. Air-dry the pellet of the magnetic beads for about 15 min.
18. Remove from the magnetic stand and add 52.5 μL of resuspension buffer. Vortex.
19. Incubate at room temperature for 2 min and then place back to the magnetic stand and wait until the liquid is clear (about 2 min).
20. Transfer 50 μL of the supernatant to the new tube.
21. Quantify the library on the High Sensitivity DNA kit on the Agilent 2100 Bioanalyzer according to the manufacturer's instructions.
22. Quantify the libraries by qPCR using the KAPA Library Quantification Kit. Add 1 mL of Illumina Primer Premix (10×) to 5 mL bottle of 2× KAPA SYBR® FAST qPCR Master Mix and mix well.
23. Dilute the sequencing library 1:1000, prepare three replicates (*see* **Note 17**).
24. Prepare the following premix per sample:
    - 12 μL of KAPA SYBR® FAST qPCR Master Mix containing Primer Premix.

- 4 μL of water.
- 4 μL of diluted sequencing library or DNA standards (there are 6 standards in the kit, they must be also used in triplicates).

25. Set the following qPCR reaction:
    - 95 °C for 5 min.
    - 35 cycles of 95 °C for 30 s and 60 °C for 45 s.
26. Analyze the results according to the manufacturer's instructions (*see* **Note 18**). In general, 10 pM of a library during cluster amplification results in ~220,000 clusters.

4 Notes

1. Using one tube of the faecal supernatant for chloroform treatment is due to the fact that the chloroform helps to remove attached phages from the bacterial cell walls. However, some phages may become damaged after the chloroform treatment, therefore the second tube with faecal supernatant remains without chloroform treatment.
2. If a visible bacterial pellet is present in the tubes after the second centrifugation, the supernatants must be transferred into new tubes and the centrifugation must be repeated.
3. The tube containing SS only, the tube containing LB only and the tube containing *E. coli* filtrate with no phage should not produce any pellet (because they do not contain any viruses).
4. This additional PEG precipitation is performed because the faecal samples contain brownish organic matter that may distort the flow cytometry analysis. An additional washing of the viral pellet with water helps to remove this organic matter.
5. This PEG precipitation is performed to remove non-binded stain molecules. It is performed also with the samples marked as "neg," because the negative controls must be processed by the same way as the stained samples.
6. The number of bacterial cells to be processed by the flow cytometry must be very low. The general rule for flow cytometry says that it is better to work with a tiny bacterial pellet than with a highly concentrated bacterial suspension. The typical flow rate is up to 5,000 cells per second, the optimal flow rate may be 1,000 cells per second.
7. Alternative nucleic acids stains may be used as well.

8. This washing with PEG may be repeated several times, if stain remains in the stained samples.
9. The settings may vary depending on the flow cytometer type. The general advice is to start with quite high fluorescence threshold and minimal particle size thresholds. By this way it will be possible to distinguish the viral particles from the electrical noise.
10. The image of a negative phage M13KE should not contain any particles, because they are below the fluorescence thresholds. The image of the stained M13KE phage should be highly fluorescent showing particles of a very small size. The reason for processing a negative phage first followed by the stained phage is that some flow cytometers do not wash themselves properly and some residual stain from the stained sample may stain the negative control. It is good to start processing negative controls before the stained samples at least during the initial experiment setting up.
11. This washing step is performed because the bacterial cells may have remained in the flow cytometer. The bacterial cells could distort the size analysis of the stained viruses that will be processed by the FC afterwards.
12. Stained SS and LB may produce a very low autofluorescence, but it should be much lower that the fluorescence produced by the positive control M13 phage.
13. If the volume of sorted sample is bigger, the volumes of the chemicals used in the following DNA extraction protocol must be adjusted accordingly. The volume 220 μL is the maximal volume of sample allowing to work with one 1.5 mL tube.
14. This dilution has been reported by Rinke et al. [17].
15. Refer to the original Nextera library preparation protocol [20] for instructions how to select the correct combination of indexes.
16. The number of cycles has been increased from the original 12 to 20 according to Rinke et al. [17].
17. The library may be diluted less times (only 1:100 or 1:10), in order to produce sufficient amplification to be quantified by the qPCR and to be comparable with library standards in the kit.
18. Refer to the KAPA Library Quantification kit manual for more details of library quantification [21].

References

1. Dutilh BE, Cassman N, McNair K et al (2014) A highly abundant bacteriophage discovered in the unknown sequences of human faecal metagenomes. Nat Commun 5:4498
2. Roux S, Enault F, Hurwitz BL et al (2015) VirSorter: mining viral signal from microbial genomic data. PeerJ 3:e985
3. Manrique P, Bolduc B, Walk ST et al (2016) Healthy human gut phageome. PNAS 113:10400–10405
4. Paez-Espino D, Eloe-Fadrosh EA, Pavlopoulos GA et al (2016) Uncovering Earth's virome. Nature 536:425–430
5. Simmonds P, Adams MJ, Benkő M et al (2017) Consensus statement: virus taxonomy in the age of metagenomics. Nat Rev Microbiol 15:161–168
6. Roux S, Hallam SJ, Woyke T et al (2015) Viral dark matter and virus–host interactions resolved from publicly available microbial genomes. elife 4:e08490
7. Deng L, Ignacio-Espinoza JC, Gregory AC et al (2014) Viral tagging reveals discrete populations in Synechococcus viral genome sequence space. Nature 513:242–245
8. Bettarel Y, Sime-Ngando T, Amblard C et al (2000) A comparison of methods for counting viruses in aquatic systems. Appl Environ Microbiol 66:2283–2289
9. Luef B, Frischkorn KR, Wrighton KC et al (2015) Diverse uncultivated ultra-small bacterial cells in groundwater. Nat Commun 6:6372
10. Legendre M, Lartigue A, Bertaux L et al (2015) In-depth study of Mollivirus sibericum, a new 30,000-y-old giant virus infecting Acanthamoeba. PNAS 112:E5327–E5335
11. Allen LZ, Ishoey T, Novotny MA et al (2011) Single virus genomics: a new tool for virus discovery. PLoS One 6:e17722
12. Martinez-Hernandez F, Fornas O, Lluesma Gomez M et al (2017) Single-virus genomics reveals hidden cosmopolitan and abundant viruses. Nat Commun 8:15892
13. Stepanauskas R, Fergusson EA et al (2017) Improved genome recovery and integrated cell-size analyses of individual uncultured microbial cells and viral particles. Nat Commun 8:84
14. Pinard R, de Winter A, Sarkis G et al (2006) Assessment of whole genome amplification-induced bias through high-throughput, massively parallel whole genome sequencing. BMC Genomics 7:216
15. Zheng Z, Advani A, Melefors O et al (2011) Titration-free 454 sequencing using Y adapters. Nat Protoc 6:1367–1376
16. Džunková M, Garcia-Garcerà M, Martínez-Priego L et al (2014) Direct sequencing from the minimal number of DNA molecules needed to fill a 454 picotiterplate. PLoS One 9:e97379
17. Rinke C, Low S, Woodcroft BJ et al (2016) Validation of picogram- and femtogram-input DNA libraries for microscale metagenomics. PeerJ 4:e2486
18. Džunková M, D'Auria G, Moya A (2015) Direct sequencing of human gut virome fractions obtained by flow cytometry. Front Microbiol 6:955
19. Thurber RV, Haynes M, Breitbart M et al (2009) Laboratory procedures to generate viral metagenomes. Nat Protoc 4:470–483
20. Illumina (2017) Nextera XT DNA library prep kit reference guide. Document #15031942v02
21. KAPAbiosystems (2017) KAPA library quantification kits for Illumina sequencing platforms. Version 4.11

Chapter 2

Tissue-Based Universal Virus Detection (TUViD-VM) Protocol for Viral Metagenomics

Claudia Kohl and Andreas Kurth

Abstract

The protocol for tissue-based universal virus detection (TUViD-VM) was developed to allow for the detection of yet unknown viruses from infected tissues. Tissues are a challenging source when it comes to virus detection. The very rapid degradation rate of tissue and the high level of host nucleic acids are the biggest challenges when examining tissues. However, to overcome these challenges, we had to decrease the amount of host nucleic acids rigorously in order to increase the amount of viral nucleic acids. We compared modern and common virological approaches to find the optimal conditions. The final TUViD-VM protocol was extensively validated by using real-time PCR and next-generation sequencing. We could increase the amount of detectable virus nucleic acids and improved the detection of viruses <75,000-fold compared with other approaches tested. The TUViD-VM protocol can be used in metagenomic and virome studies to increase the likelihood of detecting viruses from any biological source. The protocol has been widely used to detect viruses in different tissues with great success. Here we provide the bench-top protocol version of the TUViD-VM protocol.

Key words Metagenomics, NGS, Tissue, Emerging infectious diseases, Virus purification, Virome

1 Introduction

Viruses responsible for disease outbreaks in humans emerge naturally either from the human population itself or by spill-over events from animal hosts. Furthermore, viruses can also emerge unnaturally either on purpose (e.g., bioterrorist attacks) or accidentally (e.g., lab infections). Irrespective of the source of emergence, speedy identification and analysis are indispensable for taking action toward prevention and protection of the population at risk.

There are mainly five common approaches to determine the disease-causing agent in a biological sample. Either a molecular biological PCR screening, a protein-based screening with serological assays or the "open view" of electron microscopy and next-generation sequencing (NGS) can be utilized. Nevertheless, the gold standard is still in vitro virus isolation via cell culture. The

Andrés Moya and Vicente Pérez Brocal (eds.), *The Human Virome: Methods and Protocols*, Methods in Molecular Biology, vol. 1838, https://doi.org/10.1007/978-1-4939-8682-8_2, © Springer Science+Business Media, LLC, part of Springer Nature 2018

greatest disadvantage of cell-culture isolation, PCR screening, and serological approaches in terms of detecting yet unknown viruses is the lack of universality, while electron microscopy with its "open view" lacks sensitivity. However, the first four techniques are powerful tools for virus analysis, but when it comes to detection of unknown viruses they all lack sensitivity or require knowledge about the virus species in advance.

Here NGS offers "safety in numbers." Current NGS techniques such as 454 Sequencing (Roche, Mannheim, Germany), Genome Analyzer (Illumina, San Diego, CA, USA), and Ion Torrent (Thermo Fisher Scientific, Waltham, MA, USA) produce huge amounts of sequence data, hereby increasing the detection likelihood of diminutive amounts of virus compared to the host genome. The only limits seem to be how much money one would be able to spend and the capacities of calculating the results. The calculation is mainly required to distinguish between virus and host genome data to find interesting reads. But what happens when the host genome is not available? (23 genomes of 5,487 mammal species in the world; genome data is available for 0.4% of the worldwide mammalian species). For example, in the case of bats with more than 1,000 species, some of which are the most abundant mammals on the planet and carrier of highly pathogenic viruses (Ebola, Nipah, Hendra, Lyssa), only two genomes out of more than 1,000 are available.

When it comes to the detection of unknown viruses in tissue, one has to take into consideration that the amount of host-owned DNA and RNA exceeds the amount of detectable virus nucleotides many times over. To increase the possibility of detection, one approach is to first decrease the amount of host genome and then amplify the viral nucleotides to reveal a satisfying virus-to-host genome ratio. Here the complete deletion of host-owned DNA and RNA is desirable. This would not only decrease the detection limit of unknown viruses, but also save resources in NGS, as a lower amount of data would be required to deliver the same results: It is no longer necessary to have the biggest budget to deliver the biggest results.

In this study we compared theoretically and experimentally traditional and modern methods of virus purification and separation and established a protocol for the detection of viruses from any animal tissue, without prior knowledge of the virus species [1]. The number of viral reads detected per run was hereby increased to up to 50% (500,000 reads in 1,000,000 reads) instead of 0.0013% (13 reads in 1,000,000 reads). This protocol can be used in metagenomic and virome studies in NGS for the detection of viruses from any biological source, but in particular infected tissue.

2 Materials

Prepare all solutions using ultrapure water and DNase- and RNase-free analytical grade reagents. Prepare and store all reagents at room temperature (unless indicated otherwise).

2.1 Consumables

1. Autoclaved screw-cap tubes containing ceramic beads (*see* **Note 1**).
2. Sterile reaction tubes 1.5 mL.
3. 2–3/8″ PA ultracentrifuge tube suitable for SW60 rotors.
4. 3–1/2″ ultracentrifugation tube suitable for SW32Ti rotors.
5. Phase-lock gel tubes.

2.2 Solutions and Chemicals

1. Phosphate-buffered saline (PBS).
2. Sucrose solutions:
   (a) Sterile 20% sucrose w/v in PBS = 100 g (*see* **Note 2**).
   (b) 80% sucrose w/v in PBS (*see* **Note 3**).
3. TRIzol LS.
4. Chloroform–isoamyl alcohol.
5. Absolute and 70% ethanol.
6. Gel electrophoresis reagents.

2.3 Enzymes, Kits and Primers

1. Turbo DNA-Free Kit (Ambion).
2. Viral RNA Mini Kit (QIAGEN).
3. RNase-Free DNase.
4. Reverse Transcription Reagent Kit (Applied Biosystems).
5. K8N random primer, K8 primer [2].
6. dNTPs.
7. Klenow fragment and buffer.
8. PCR product purification kit.
9. AmpliTaq Gold™ DNA Polymerase with Buffer II and $MgCl_2$.

2.4 Centrifuges, Rotors and Shakers

1. Ultracentrifuge (125,000 × *g*).
2. Tissue honogenizer.
3. Vortex.
4. Bench-top centrifuge (425–15,000 × *g*).
5. SW60 rotor and brackets.
6. SW32Ti rotor and brackets.
7. Scale for taring ultracentrifugation tubes.
8. PCR cycler.
9. Gel electrophoresis chamber and power block.

3 Methods

Carry out all procedures at room temperature unless otherwise specified. To allow for throughout examination of the sample, it should be snap-frozen in liquid nitrogen after necropsy. However, we found that *RNAlater®* storage at room temperature could be used, alternatively [3]. All procedures are to be carried out in a biosafety cabinet. The necessary biosafety level is dependent on the sample and is determined through a risk analysis.

3.1 Tissue Homogenate

1. Cut a small cube of tissue (0.5–1 cm^3) and place into an autoclaved screw-cap tube containing 1 mL of phosphate-buffered saline (PBS) buffer and 20–30 sterile ceramic beads.
2. Disrupt tissue by shaking four times at maximum speed at intervals of 15 s by using the FastPrep-24 Instrument. Place tube on ice during the intervals to avoid RNA degradation.
3. The duration of this procedure is ≈0.5 h. (*see* **Note 4**).

3.2 Clearing Centrifugation

1. Place 200 μL of homogenate into a 1.5 mL tube and vortex vigorously.
2. Centrifuge the homogenate at 425 × *g* for 5 min in a benchtop centrifuge.
3. Transfer the supernatant (≈170 μL) into a clean tube and discard the pellet.
4. The duration of this procedure is ≈0.25 h.

3.3 Ultracentrifugation for Virus Particle Separation

1. Pipet 250 μL of 80% (wt/vol) sucrose solution into a 2–3/8″ PA ultracentrifuge tube.
2. Gently overlay with ≈3 mL of 20% (wt/vol) sucrose solution. The visibility of the phase interface between the 80% and 20% sucrose solutions should be checked. When the layers are clearly separated go to the next step.
3. Overlay the sucrose solution gently with cleared tissue supernatant from **step 3** in Subheading 3.2 and add PBS to adjust the volume and tare.
4. Centrifuge in an SW60 rotor at 120,000 × *g* at 4 °C for 2 h.
5. The duration of this procedure is ≈2 h.

3.4 Ultracentrifugation to Pellet Virus Particles

1. Collect the thin layer on the interface between the 20% and 80% sucrose solutions and transfer it into a fresh 3–1/2″ tube.
2. Resuspend the collected interphase in ≈40 mL of PBS and mix gently by pipetting up and down with a wide serological pipette.
3. Centrifuge the suspension at 125,000 × *g* and 4 °C for 1 h in a SW32 rotor. Discard the supernatant.
4. The duration of this procedure is ≈1 h (*see* **Note 5**).

3.5 DNA Digestion

1. Resuspend the pellet from **step 3** in Subheading 3.4 in 245 μL of 1× digestion buffer (Turbo DNA Free Kit).
2. Add a total of 5 μL of Turbo DNase and incubate at 37 °C for 30 min.
3. Transfer the suspension to a fresh 1.5 mL reaction tube.
4. Add a total of 10 μL of stop reagent, incubate at room temperature for 1 min, and centrifuge at 425 × *g* for 3 min.
5. Transfer the supernatant to a fresh tube and discard the pellet.
6. The duration of this procedure is ≈0.75 h.

3.6 Combined TRIzol LS RNA Extraction (See Note 6)

1. Add a total of 750 μL of TRIzol LS to ≈250 μL of supernatant from the previously described procedures and homogenize by pipetting up and down 10 times.
2. Incubate the mixture at room temperature for 5 min and centrifuge at 15,000 × *g* for 10 min.
3. Transfer the supernatant to a precentrifuged phase-lock gel tube.
4. Add a total of 200 μL of chloroform–isoamyl alcohol and mix by inverting the tube vigorously.
5. Incubate the tube at room temperature for 15 min and centrifuge at 15,000 × *g* for 15 min.
6. Transfer approximately 280 μL of supernatant from the phase-lock gel tube to a fresh tube containing 1120 μL of AVL lysis buffer without carrier RNA.
7. Add a total of 700 μL of absolute ethanol and mix by pulse vortexing.
8. Transfer the solution in 600 μL portions to a QIAamp Mini Column and centrifuge at 7000 × *g* for 1 min, discard the filtrate. Place the column in a new collection tube, load again and centrifuge until the complete lysate is added to the column.
9. Add a total of 500 μL of 70% (wt/vol) ethanol to the column and centrifuge at 7000 × *g* for 3 min.
10. Add a mixture of 10 μL of DNase and 70 μL of RDD buffer (RNase-Free DNase Set) to the column and incubate at room temperature for 15 min, as described by the manufacturer.
11. Wash the column with 500 μL of AW1 buffer, centrifuge at 7000 × *g* for 1 min, and discard the filtrate.
12. Place the column in a new tube, add 500 μL of AW2 buffer, centrifuge the tube at maximum speed for 3 min, and discard the filtrate.
13. Place the column in a new tube and centrifuge the tube at maximum speed for 1 min to dry the column.

14. Add a total of 30 μL of elution buffer to the column, incubate at room temperature for 5 min, and centrifuge the column in a new 1.5 mL tube. Again, add a total of 30 μL of elution buffer to the column, incubate at room temperature for 5 min, and centrifuge in the same tube.
15. Chill the RNA (≈60 μL) on ice. The duration of this procedure is ≈3 h.

3.7 Random Amplification

1. cDNA synthesis: Store 30 μL of the freshly isolated RNA at −80 °C as a backup. Use the remaining 30 μL RNA for the cDNA synthesis. Produce cDNA utilizing the Applied Biosystems™ TaqMan™ Reverse Transcription Reagents Kit and adapt the volumes for a 50 μL reaction containing 30 μL of RNA. Add 2 μL of K8N random primers and 3.2 μL of dNTPs directly to the 30 μL of RNA and heat to 95 °C for 5 min before quenching on ice. Add the remaining contents of the master mix and heat to 42 °C for 60 min before inactivating the enzymes at 95 °C for 10 min.

   The reaction mix for cDNA synthesis, adapted to an RNA volume of 30 μL, consists of 30.0 μL of RNA from **step 1** in Subheading 3.7; 2.0 μL of 40 μM K8N primers; 3.2 μL of 25 mM dNTPs; 4.0 μL of buffer; 9.0 μL of $MgCl_2$; 0.8 μL of RNase inhibitor; 0.6 μL of reverse transcriptase; and 0.6 μL of H_2O.
2. Double-stranded cDNA is produced by mixing 2 μL of K8N random primers, 3 μL of Klenow buffer and 2 μL of 2.5 mmol/L dNTPs with 19 μL of cDNA from **step 2** of Subheading 3.7. The mixture is heated at 95 °C for 2 min and cooled to 4 °C. Add a total of 1.67 μL of Klenow fragment and heat the mixture at 37 °C for 60 min.
3. The double-stranded cDNA should be purified by using the MSB Spin PCRapace Purification Kit (Invitek) and an elution volume of 30 μL.
4. Random amplification is performed by using the procedures reported by Stang and Korn [2]. The reaction mix for random amplification consists of: 13 μL of H_2O; 5 μL of AmpliTaq Gold Buffer; 4 μL of 2.5 mM dNTPs; 10 μL of 25 mM $MgCl_2$; 4 μL of 10 μM K-Primer (not K8N!); 9 μL of 5 U/μL AmpliTaq Gold; and 5 μL of dscDNA (**step 3** in Subheading 3.7).

   The cycling conditions for random amplification, also adapted from Stang and Korn [2], are as follows: a denaturation cycle at 95 °C for 5 min; followed by 5 cycles at 95 °C for 1 min, 59 °C for 1 min and 72 °C for 1 min; then 33 cycles at 95 °C for 20 s, 59 °C for 20 s, and 72 °C for 1 min; and a final extension at 72 °C for 7 min; and a final hold at 4 °C for approx. 1 h 45 min.

5. The random amplification product should appear as a smear in between 200 and 2000 bp and can be visualized by agarose gel electrophoresis. The size of interest (mainly between 500 and 1000 bp) can be extracted via gel purification kits.
6. The duration of this procedure is ≈4.5 h. Sequence information can be obtained by either cloning into sequencing vectors (*see* **Note 7**) or by NGS (*see* **Note 8**).

4 Notes

1. *Autoclaved screw-cap tubes containing ceramic beads.* We used the in-house autoclaved screw-cap tubes containing ceramic beads. But there are some companies from which you can buy ready-to-use tubes with beads specified for distinct tissues (e.g., mosquitoes, liver and brain).
2. *Sucrose solution 20%.* Weigh 20% sucrose (100 g in 500 mL) within a measuring glass (1000 mL) and fill with PBS to a volume of 500 mL. Mix with a magnetic stirrer until all sucrose crystals have dissolved (approx. 15 min). Use a water-jet vacuum pump for sterile filtration of the sucrose solution into a fresh and sterile 500 mL glass flask (sterile one-way 0.22 μm filter). Keep at room temperature to avoid crystallization. Solution should be usable for several weeks.
3. *Sucrose solution 80%.* Weigh 80% sucrose (80 g in 100 mL) in a measuring glass (250 mL) and fill with PBS to a volume of 100 mL. Mix with a magnetic stirrer until all sucrose crystals have dissolved (approx. 120 min). Fill into a small sterile glass bottle. Sterile filtration of the solution is not possible due to its high viscosity. Remove magnetic stirrer from the solution and keep at room temperature to avoid any crystallization. Solution should be usable for several weeks.
4. *Storage of the homogenate.* The tissue homogenate should be used straight away to avoid any degradation. If storage is necessary, the homogenate can be stored at −80 °C.
5. *Virus precipitation using Peg-It.* As an alternative method, virus particles can be precipitated in the refrigerator overnight by using the Peg-It solution (System Biosciences). This was not tested for RNAlater®-stabilized samples.
6. *RNA extraction using Direct-zol™.* We have tested an alternative extraction using the Direct-zol™ RNA MiniPrep Plus Kit (ZYMO research). This kit is much quicker than the described method, but is a little less effective. We have found a difference of 1–2 Ct values when using Direct-zol™ instead

of Trizol in combination with the Viral RNA Mini Kit from Qiagen. But when time is limited, Direct-zol™ may be a good solution. This was also tested with the *RNAlater®*-stabilized samples.

7. *Molecular cloning*. This is an easy and cheap method to analyze the purified sequences. We have used sequencing vectors like the TOPO® TA Cloning kits (Thermo Fisher) that work on the basis of TA tagging. This means if a polymerase (like the Ampli Taq gold) is used that produces 3′-A overhangs, the PCR product can be cloned directly into a vector system with T overhangs. It is best to use fresh PCR products; however, if this is not possible the PCR products can be refreshed by adding polymerase and DNTPs and heating for elongation for approx. 5 min. We used Top10 chemical competent cells for transformation.

   The random products are often complex and different; therefore an overnight ligation at 4 °C will increase the amount of positive clones. The clones were evaluated by simple blue-white screening and amplified using the M13 primers with the protocol described by the manufacturer. We did not grow clones for plasmid production, but just touched a clone with a sterile pipette tip and dipped it into the master mix. An agarose gel electrophoresis showed which clones had an insert; these were picked for Sanger sequencing using again the M13 primers. With this approach a few hundred clones can easily be analyzed.

8. *Next-generation sequencing*. In the original protocol the samples were analyzed by using the Ion Torrent. Therefore, the protocol is described here. However, this protocol works also well with Illumina HiSeq and MiSeq. RNA samples were fragmented by using the NEBNext Magnesium RNA Fragmentation Module (New England Biolabs). RNA was purified by using RNeasy MinElute (QIAGEN). For cDNA synthesis, Superscript II and Murine RNAse inhibitor (New England Biolabs) were used. Second-strand synthesis was performed by using the NEBNext mRNA Second Strand Synthesis Module (New England Biolabs) and purified by using the MinElute PCR Purification Kit (QIAGEN). Double-stranded cDNA, DNA, and random PCR products were quantified by using the Qubit HS dsDNA Kit (Invitrogen Life Technologies). Sequencing libraries were established by Ion Xpress Plus Fragment Library Kit (without chemical fragmentation) with indices (Ion Xpress Barcode Adapters 1–16 Kit). The sequencing library was then amplified by using an emulsion-based clonal amplification PCR in the Ion OneTouch 200 Template v2 DL Kit and enriched by using an Ion OneTouch Enrichment System. Sequencing was performed on an Ion Torrent PGM in the Ion PGM Sequencing 300 Kit with the Ion 318 Chip Kit (Invitrogen Life Technologies).

Acknowledgments

The authors are grateful to Ursula Erikli for copy-editing.

References

1. Kohl C, Brinkmann A, Dabrowski PW et al (2015) Protocol for metagenomic virus detection in clinical specimens. Emerg Infect Dis 21(1):48–57
2. Stang A, Korn K, Wildner O et al (2005) Characterization of virus isolates by particle-associated nucleic acid PCR. J Clin Microbiol 43(2):716–720
3. Kohl C, Wegener M, Nitsche A et al (2017) Use of RNALater® preservation for Virome sequencing in outbreak settings. Front Microbiol 8:1888

Chapter 3

Protocol for Generating Infectious RNA Viromes from Complex Biological Samples

Sonia Monteil-Bouchard, Sarah Temmam, and Christelle Desnues

Abstract

This chapter proposes a simple, standardized protocol for generating RNA viromes from complex host-associated biological samples of various origins. Compared to other existing protocols to generate RNA viromes, this protocol preserves the infectivity of viral particles and allows for downstream applications such as viral characterization and isolation tests.

Key words Virome, Arthropods, Metagenomics, RNA viruses, Sucrose density gradient, Viral zoonosis

1 Introduction

RNA viruses are responsible for many human and animal infectious diseases and most (emerging) zoonoses are caused by RNA viruses that have crossed barrier species and jumped from animal to human host [1]. A better knowledge of the viruses that are present in potential animal vectors/reservoirs, as well as a rapid detection of RNA virus infection, is thus critical for the diagnosis, treatment, control, and prevention of human and animal diseases. Nucleic acid-based technologies such as Polymerase Chain Reaction (PCR) using targeted primers or probes have facilitated the detection and the characterization of viruses in complex environmental, animal, or human samples. However, these require previous knowledge on the viral genomes and thus can largely miss divergent or novel species. This issue is particularly challenging because RNA viruses usually present great genetic diversity as a consequence of high mutation rates and recombination events. In the last decade, shotgun high-throughput sequencing of the viral genetic material protected by the capsid shell (aka viral metagenomics) has emerged as a promising tool to assess the viral composition of a particular sample. Since then, viral metagenomics has been widely used to study the diversity of DNA and RNA viruses

Andrés Moya and Vicente Pérez Brocal (eds.), *The Human Virome: Methods and Protocols*, Methods in Molecular Biology, vol. 1838, https://doi.org/10.1007/978-1-4939-8682-8_3, © Springer Science+Business Media, LLC, part of Springer Nature 2018

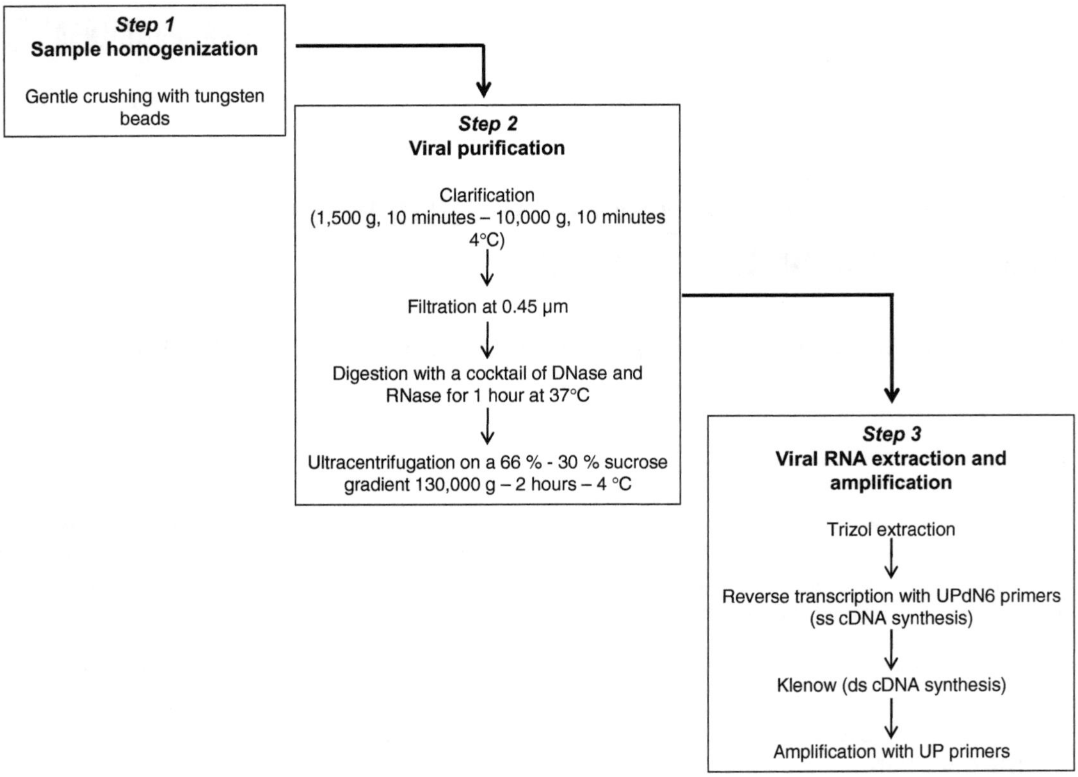

Fig. 1 Overview of the general procedure

from the environment [2–6] and to identify the causative viral agent in clinical/veterinary diagnostic virology [7–12].

One of the main limitations of viral metagenomics in complex host-associated samples is that the ratio between viral and non-viral (bacterial and host) sequences is usually low. To increase the signal-to-noise ratio, several pretreatments to purify and enrich viral particles such as homogenization, filtration, centrifugation, nuclease treatment, and density gradient ultracentrifugation can be applied to the samples before nucleic acids extraction and sequencing [13–21]. In this chapter, we proposed a simple and rapid protocol for generating RNA viromes from complex biological samples, which efficiently diminishes non-viral nucleic acid contamination for NGS sequencing and preserves the infectivity of viral particles for further isolation assays. This protocol has been validated on human body lice engorged on human blood which has been artificially contaminated with RNA viruses [22]. The general procedure includes three major steps (Fig. 1): (1) Homogenization, (2) Viral purification, (3) Viral nucleic acid extraction and amplification. First, the sample is gently homoge-

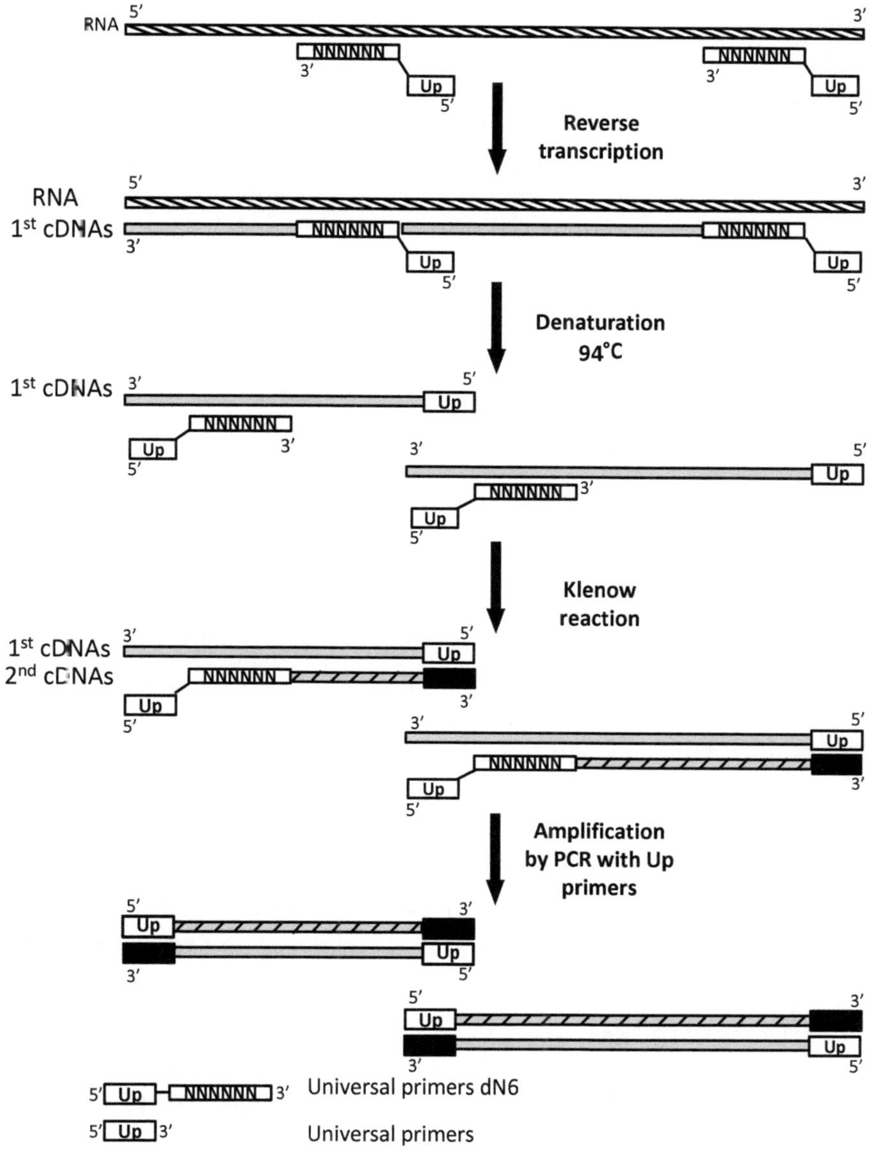

Fig. 2 Mechanism of the reverse transcription and sequence-independent amplification (SIA). Modified from [23]

nized using tungsten beads for optimizing viral particle release (step 1). Viral particles in suspension are then purified from the sample by clarification, filtration, digestion with nucleases, and a sucrose density gradient ultracentrifugation (step 2). Finally (step 3), viral RNAs are extracted, reverse transcribed, and sequence-independently amplified (SIA) using a hexamer tagged with a 26 nucleotide primers as described previously [23] (Fig. 2).

2 Materials

2.1 Homogenization

1. 1× Phosphate-Buffered Saline (PBS): Dissolve 0.2 g of KCl, 0.24 g of KH_2PO_4, 8 g of NaCl, and 1.44 g of Na_2HPO_4-$7H_2O$ in 800 mL of sterile water. Adjust pH to 7.4 with HCl. Adjust volume to 1 L with sterile water and filter at 0.02 μm. Aliquot and store at +4 °C.
2. Sterile scalpel.
3. Sterile tweezer.
4. 3 mm tungsten beads.
5. Tissue Lyser.
6. 70% ethanol.

2.2 Viral Particle Purification

1. Bench centrifuge.
2. 0.45 μm PVDF filter 13 mm.
3. 70% ethanol.
4. 2 U/μL Turbo DNase, stored at −20 °C.
5. 1× Turbo DNase Buffer (buffer composition not specified by the supplier, Ambion), supplied as 10× stock solution, stored at −20 °C.
6. 120 U/mg RNase A supplied lyophilized. Dissolve 50 mg in 1.7 mL of sterile DNase/RNase-free water to obtain a stock solution at 3.6 U/μL. Aliquot and store at −20 °C.
7. 20 U/μL Exonuclease I, stored at −20 °C (*see* **Note 1**).
8. 25 U/μL Benzonase, stored at −20 °C (*see* **Note 2**).
9. 1× PBS sterile, 0.02 μm filtered.
10. 66% sucrose solution: weigh 6.6 g of sucrose under a class II microbiological safety cabinet; add 0.02 μm-filtered sterile 1× PBS to a volume of 10 mL (*see* **Note 3**). Prepare fresh or store at 4 °C for a maximum of 3 days.
11. 30% sucrose solution: weigh 6 g of sucrose under a class II microbiological safety cabinet; add 0.02 μm-filtered sterile 1× PBS to a volume of 20 mL. Prepare fresh or store at 4 °C for a maximum of 3 days.
12. Anotop™ 25 filters, inorganic membrane 0.02 μm filters.
13. Rotor SW55 Ti.
14. Autoclaved swinging buckets.
15. Ultracentrifuge.
16. Tubes, Thinwall, Ultra-Clear™, 5 mL.

2.3 Viral RNA Extraction

1. Trizol LS.
2. Chloroform molecular biology grade.

3. Chemical hood.
4. RNA Clean and Concentrator™-25 kit (*see* **Note 4**).
5. 40 U/μL RNaseOUT™, stored at −20 °C.
6. Refrigerated bench centrifuge.

2.4 RNA Processing

1. PCR UV cabinet.
2. DNAZap™ (a reagent for DNA/RNA decontamination).
3. 10 μM dNTPs.
4. 10 μM UPdN6 primers: 5'-GCCGGAGCTCTGCAGAATT CNNNNNN-3′.
5. 10 μM Up primer: 5'-GCCGGAGCTCTGCAGAATTC-3′.
6. 200 U/μL Superscript III reverse transcriptase, stored at −20 °C.
7. 1× First strand, supplied as 5× stock solution buffer [250 mM Tris–HCl, pH 8.3, 375 mM KCl, 15 mM MgCl] and stored at −20 °C.
8. 40 U/μL RNaseOUT™, stored at −20 °C.
9. 5 U/μL DNA polymerase I Klenow fragment, stored at −20 °C.
10. 1× NEBuffer™ 2 [50 mM NaCl, 10 mM Tris–HCl, 10 mM $MgCl_2$,1 mM DTT, pH 7.9], supplied as 10× stock solution and stored at −20 °C.
11. Magnetic Particle Concentrator (MPC).
12. Agencourt™ AMPure™ XP beads.
13. Freshly prepared 70% ethanol.
14. 2.5 U/μL LongAmp™ Taq DNA polymerase, stored at −20 °C.
15. 1× LongAmp™ Taq reaction buffer [60 mM Tris-SO_4, 20 mM $(NH_4)_2SO_4$, 2 mM $MgSO_4$, 3% Glycerol, 0.06% IGEPAL®CA-630, 0.05% Tween®20, pH 9.1], supplied as 5× stock solution and stored at −20 °C.
16. Sterile DNase/RNase-free water.
17. Thermocycler.

2.5 Purification of SIA

1. NucleoFast™ 96 PCR.
2. Waste collection plate.
3. 1× Tris-EDTA (TE) buffer: 10 mM Tris–HCl, 1 mM EDTA, sterile and filtered at 0.02 μm. Store at room temperature (RT).
4. Bench centrifuge.

3 Methods

For all the procedures, a negative control consisting of 0.02 μm-filtered sterile 1× PBS is processed and sequenced in parallel with the sample.

Homogenization and viral purification must be performed in a class II microbiological safety cabinet and disposable material (tips, tubes, pipettes, etc.) must be decontaminated before being placed in an appropriate waste container.

3.1 Homogenization

1. For each sample, sterilize two 3 mm-tungsten beads in 70% ethanol for 20 min in a Petri dish.
2. Rinse the beads in another Petri dish with 0.02 μm-filtered sterile 1× PBS.
3. Add the sample and the beads to the 2 mL tube.
4. Add 2 mL (the volume has to be adjusted according to the size of the sample) of 0.02 μm-filtered sterile 1× PBS to the 2 mL tubes. The proposed 2 mL is used, for instance, for a sample of 20 body lice (*Pediculus humanus*) or 20 mosquitoes (*Aedes albopictus*).
5. Crush for 2 min at 25 Hz in a Tissue Lyser.
6. Another run can be performed if any remaining hard pieces of the sample are observed in the tube.

3.2 Viral Purification

1. Sterilize the Ultra-Clear™ tubes for 20 min with 70% ethanol in a class II microbiological safety cabinet and rinse with 0.02 μm-filtered sterile 1× PBS. Remove all remaining drops by drying in a class II microbiological safety cabinet.
2. Clarify the homogenized sample at 1,500 × *g* for 10 min at 4 °C (*see* **Note 5**).
3. Collect the supernatant and clarify again at 10,000 × *g* for 15 min at 4 °C (*see* **Note 6**).
4. Pre-wet a 0.45 μm-filter with 0.02 μm-filtered sterile 1× PBS, then slowly filtrate the supernatant at 0.45 μm.
5. Digest the filtrate with a cocktail of DNase and RNase. Add: 1× of Turbo DNase buffer, 20 U of Turbo DNase, 25 U of RNase A, 20 U of exonuclease I and 25 U of benzonase.
6. Incubate for 1 h at 37 °C in a water bath.
7. Pour every sucrose density with a pipet (Fig. 3). Begin with 800 μL of the 66% sucrose layer (Fig. 3a, b) (*see* **Note 7**). Add 2.7 mL of the 30% sucrose layer, very slowly so as not to disturb the boundary before centrifugation (Fig. 3c–e).
8. Very slowly and carefully add 1.3 mL of sample with a pipet (Fig. 3f, g). Avoid disturbing the gradient.
9. Verify that the Ultra-Clear™ tubes are filled until the last 5 mm from the top. If this is not the case, complete the volume with 0.02 μm-filtered sterile 1× PBS (*see* **Note 8**).
10. Equilibrate the tubes (using an analytical balance) with 0.02 μm-filtered sterile 1× PBS.

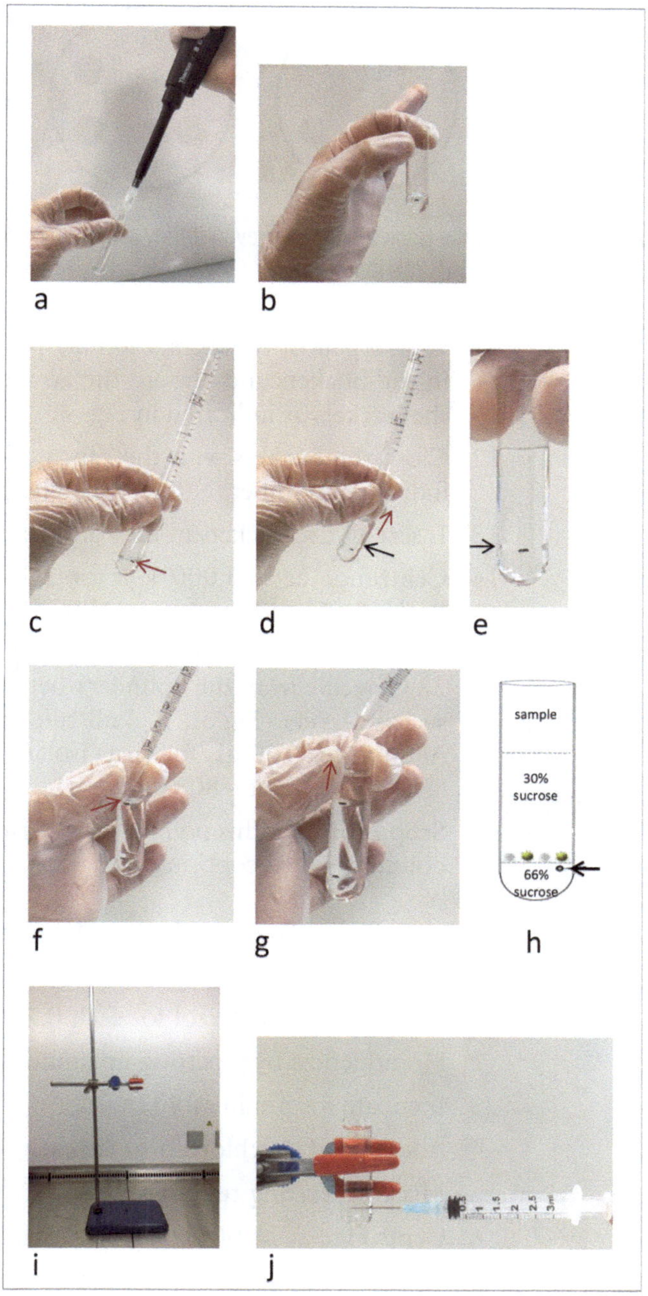

Fig. 3 Detailed method for pouring the sucrose gradient and harvesting viral fraction. A first layer consisting of 800 μL of 66% sucrose is deposited in the tube (**a**). Put a mark with a permanent marker for delimiting the two layers (**b**). Tilt the tube and place the pipette on the tube wall just above the 66% layer (**c**, red arrow) (do not touch the 66% layer) and pour the 30% sucrose layer very slowly. Move up the pipette while you are pouring the second layer (red arrow **d**). The presence of a visible boundary between the two layers (black arrow **d** and **e**) is a good indicator of a successful gradient pouring. For the sample, process as above for the 30% sucrose layer (**f**, **g**): place the pipette on the tube wall just above the 30% layer (**f**, red arrow), pour the sample while moving the pipette up (**g**, red arrow). After ultracentrifugation, viral particles are concentrated at the boundary between the 66% and the 30% sucrose layer (**h**). Hang the tube tightly onto the bracket (**i**) and harvest the viral fraction with a 2 mL syringe equipped with a 23 G needle, by piercing the tube just below the mark separating the two sucrose layers (**j** and **h** black arrow)

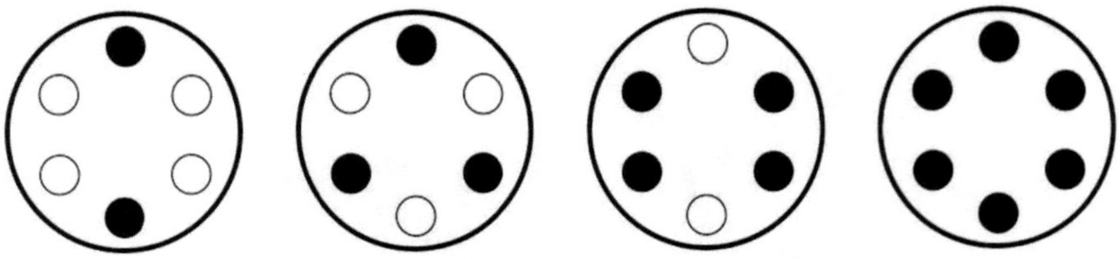

Fig. 4 Bucket positions on the rotor. Hook all buckets, even if they are empty. Arrange the buckets in the rotor as shown. Modified from the Sw55Ti manual

11. With a pair of tweezers, carefully put the Ultra-Clear™ tubes in the buckets and choose the arrangement for equilibrating the buckets as indicated in Fig. 4.
12. Close the buckets with the cap. Each bucket and cap has an individual number. Number matching should be respected.
13. Insert all the buckets in the rotor (empty and loaded).
14. Centrifuge at 130,000 × *g* for 2 h at 4 °C in a Sw 55 Ti Beckman-Coulter rotor (*see* **Note 9**).
15. Harvest the viral fraction with a 2 mL syringe equipped with a 23 G needle from the boundary between the 66% and the 30% sucrose layers (Fig.3i, j). Pull out approximately 0.8–1 mL. Aliquot 2 × 100 μL for virus isolation assays on adapted cell lines and store at −80 °C.
16. **Stop point**. At this point it is possible to freeze at −80 °C the boundary or to continue with Trizol extraction.

3.3 Viral RNA Extraction

Extraction with Trizol requires a chemical hood equipped with a filter adapted for organic vapors.

1. Add 0.75 mL of Trizol for each 0.25 mL of sample.
2. Homogenize by pipetting up and down several times.
3. Incubate for 5 min at RT.
4. Add 0.2 mL of chloroform for each 0.75 mL of Trizol.
5. Mix by inverting tubes for 15 s.
6. Centrifuge at 12,000 × *g* for 15 min at 4 °C (*see* **Note 10**).
7. Pipette out the aqueous phase, and transfer it into a new 15 mL tube. After this step, a kit is used to purify the extracted viral RNAs.
8. Add one volume of 100% ethanol to the aqueous phase and mix by inverting tube.
9. Load 800 μL of the sample onto the column, and centrifuge at 10,000 × *g* for 30 s. If the sample is >800 μL, the column can be reloaded.
10. Wash the column with 400 μL of RNA Prep buffer, and centrifuge at 10,000 × *g* for 30 s.

11. Wash the column with 700 μL of RNA Wash buffer, and centrifuge at 10,000 × *g* for 30 s.
12. Wash the column with 400 μL of RNA Wash buffer and centrifuge at 10,000 × *g* for 2 min.
13. Transfer the column into as new RNase-free tube.
14. Elute with 25 μL of DNase/RNase-free water.
15. Add 40 U of RNaseOUT™.
16. After the elution step, you can continue with the reverse transcription step or aliquot the viral RNA and freeze it at −80 °C.

3.4 RNA Processing

1. Clean the pipettes, bench and rack with DNAZap™ under the PCR UV cabinet.
2. Put 10 μL of viral RNAs in PCR strip tubes and denature them at 65 °C for 5 min. Cool immediately on ice (*see* **Note 11**).
3. Prepare the reverse-transcription mix under the PCR UV cabinet by adding 200 U of Superscript III, 1× of first strand buffer, 20 U of RNase OUT, 0.625 mM of dNTPs, 0.94 μM of UPdN6 primers and complete the volume to 20 μL with sterile DNase/RNase-free water. Prepare the mix on ice or in a cooler (*see* **Note 12**).
4. On ice or in a cooler, add 10 μL of the mix to the denatured viral RNAs, and mix slowly by pipetting up and down.
5. Incubate at 25 °C for 5 min −35 °C, 15 min −55 °C, 30 min −94 °C for 2 min in a thermocycler and cool immediately on ice or in a cooler.
6. **Stop point.** It is possible at this point to freeze the single-stranded cDNA (ss cDNA) at −20 °C, in this case you must denature the ss cDNA at 94 °C for 2 min before continuing to **step** 7.
7. For the synthesis of the second-strand, prepare a mix containing 1× NEBuffer™ 2, 8 U DNA polymerase I Klenow fragment, 0.33 mM dNTPs and complete at 30 μL with DNase/RNase-free sterile water. Prepare the mix on ice or in a cooler (*see* **Note 12**).
8. Add 10 μL of the mix on the denatured ss cDNA (20 μL) and mix slowly by pipetting up and down.
9. Incubate at 37 °C for 1 h.
10. **Stop point.** It is possible at this point to freeze the double-stranded cDNAs (ds cDNA) at −20 °C, or to continue with Agencourt™ AMPure™ XP beads purification.
11. Adjust the volume of the sample to 50 μL with 0.02 μm-filtered 1× TE.
12. Vortex the Agencourt™ AMPure™ XP beads bottle for 20 s or until the beads are completely resuspended.

13. Add 46 μL of Agencourt™ AMPure™ XP beads, and mix 10 times by pipetting up and down. Incubate for 5 min at RT.
14. Place the tube on the Magnetic Particle Collector (MPC) to separate the beads from the solution. Wait for the solution to clear (it may take up to 3–4 min).
15. Pipette the cleared supernatant and discard. This step must be performed while the tube is positioned on the MPC.
16. Wash the pellet of beads with 300 μL of freshly prepared 70% ethanol. This step must be performed while the tube is positioned on the MPC. Pipette out the 70% ethanol and discard. Repeat this step once.
17. Keep the tubes on the MPC under the hood until the pellet is completely dried (about 5–10 min but do not let the pellet over dry) (*see* **Note 13**).
18. Remove the tubes from the MPC and add 30 μL of 1× TE to the tube and mix by pipetting up and down 10 times then incubate for 5 min at RT.
19. Place the tube on the MPC for 1 min to separate the beads from the supernatant containing the ds cDNAs.
20. Transfer 28 μL of the supernatant into a new tube.
21. **Stop point**. It is possible at this point to freeze the ds cDNAs at −20 °C, or to continue with amplification.
22. Amplify the ds cDNAs in a mix containing 1× LongAmp™ Taq reaction buffer, 2.5 U LongAmp™ Taq DNA polymerase, 0.2 mM dNTPs, 1 μM UP primers and complete at 25 μL with DNase/RNase-free sterile water. Prepare the mix on ice or in a cooler (*see* **Note 12**).
23. Dispense 17 μL of the amplification mix to each PCR strip tubes.
24. Add 8 μL of ds cDNAs.
25. Incubate at 94 °C for 1 min, then 40 cycles at 55 °C for 1 min, and next at 72 °C for 3 min, and finally at 72 °C for 7 min for final extension. Hold at 10 °C in a thermocycler.

3.5 Purification of SIA

1. Adjust the sample volume to 100 μL with 0.02 μm-filtered sterile 1× TE.
2. Transfer the sample to the NucleoFast™ 96 PCR.
3. Place a waste collection plate under the NucleoFast™ 96 PCR.
4. Centrifuge at 2200 × *g* for 15 min or until the well is empty of any liquid.
5. Wash the membrane with 100 μL of 0.02 μm-filtered sterile 1× TE.
6. Centrifuge at 2200 × *g* for 15 min or until the well is empty of any liquid.

7. Add 40 μL of 0.02 μm-filtered sterile 1× TE.
8. Incubate for 5 min at RT.
9. Mixing by pipetting up and down five times and transfer the sample into a new 1.5 mL tube.

4 Notes

1. This enzyme degrades single-stranded DNA.
2. This enzyme degrades double- and single-stranded, circular and linear RNA and DNA.
3. The 66% sucrose solution may require heating for 10 min at 55 °C for correct dissolution. Once all the sucrose is correctly dissolved, the volume can be adjusted with 0.02 μm-filtered sterile 1× PBS.
4. This kit allows the purification of RNAs from the aqueous phase after Trizol extraction.
5. This speed allows pelleting cellular debris.
6. This speed allows pelleting mitochondria, lysosomes, peroxisomes and chloroplasts.
7. Use a permanent marker to delimit the two sucrose layers.
8. If the tubes are not full, they can collapse during the ultracentrifugation.
9. After ultracentrifugation, great care should be taken to remove and manipulate the tubes from the buckets for not disturbing the gradient.
10. After centrifugation, three phases are obtained: the upper, colorless phase (which is the aqueous phase containing RNA), the interphase and the lower phase.
11. Add a negative control (10 μL of sterile water DNase/RNase free) that will do the three steps of the RNA processing: reverse transcription, Klenow and amplification by PCR.
12. For pipetting errors, prepare one supplementary reaction.
13. At the beginning of the drying the pellet is bright. When it turns mat the TE can be added. Do not over dry the pellet (it should not appear cracked) otherwise the elution efficiency can decrease.

Acknowledgments

This work was supported by a grant from the Agence Nationale de la Recherche number ANR-13-JSV6-0004 to CD.

References

1. Jones KE, Patel NG, Levy MA et al (2008) Global trends in emerging infectious diseases. Nature 451:990–993
2. Angly FE, Felts B, Breitbart M et al (2006) The marine viromes of four oceanic regions. PLoS Biol 4:e368
3. Djikeng A, Kuzmickas R, Anderson NG et al (2009) Metagenomic analysis of RNA viruses in a fresh water lake. PLoS One 4(9):e7264
4. Fancello L, Trape S, Robert C et al (2013) Viruses in the desert: a metagenomic survey of viral communities in four perennial ponds of the Mauritanian Sahara. ISME J 7:359–369
5. Rosario K, Breitbart M (2011) Exploring the viral world through metagenomics. Curr Opin Virol 1:289–297
6. Whon TW, Kim MS, Roh SW et al (2012) Metagenomic characterization of airborne viral DNA diversity in the near-surface atmosphere. J Virol 86:8221–8231
7. Barzon L, Lavezzo E, Militello V et al (2011) Applications of next-generation sequencing technologies to diagnostic virology. Int J Mol Sci 12:7861–7884
8. Belák S, Karlsson OE, Blomström AL et al (2013) New viruses in veterinary medicine, detected by metagenomic approaches. Vet Microbiol 165:95–101
9. Blomström AL (2011) Viral metagenomics as an emerging and powerful tool in veterinary medicine. Vet Q 31(3):107–114
10. Capobianchi MR, Giombini E (2013) Rozera G (2013) next-generation sequencing technology in clinical virology. Clin Microbiol Infect 19:15–22
11. Chiu CY (2013) Viral pathogen discovery. Curr Opin Microbiol 16:468–478
12. Temmam S, Davoust B, Berenger JM et al (2014) Viral metagenomics on animals as a tool for the detection of zoonoses prior to human infection? Int J Mol Sci 15:10377–10397
13. Conceição-Neto N, Zeller M, Lefrère H et al (2015) Modular approach to customise sample preparation procedures for viral metagenomics: a reproducible protocol for virome analysis. Sci Rep 5:16532
14. Hall RJ, Wang J, Todd AK et al (2014) Evaluation of rapid and simple techniques for the enrichment of viruses prior to metagenomic virus discovery. J Virol Methods 195:194–204
15. Kohl C, Brinkmann A, Dabrowski PW et al (2015) Protocol for metagenomic virus detection in clinical specimens. Emerg Infect Dis 21(1):48–57
16. Lewandowska DW, Zagordi O, Geissberger F-D et al (2017) Optimization and validation of sample preparation for metagenomic sequencing of viruses in clinical samples. Microbiome 5:94
17. Li L, Deng X, Mee ET et al (2015) Comparing viral metagenomics methods using a highly multiplexed human viral pathogens reagent. J Virol Methods 213:139–146
18. Manso CF, Bibby DF, Mbisa JL (2017) Efficient and unbiased metagenomic recovery of RNA virus genomes from human plasma samples. Sci Rep 7:4173
19. Rosseel T, Ozhelvaci O, Freimanis G et al (2015) Evaluation of convenient pretreatment protocols for RNA virus metagenomics in serum and tissue samples. J Virol Methods 222:72–80
20. Thurber RV, Haynes M, Breitbart M et al (2009) Laboratory procedures to generate viral metagenomes. Nat Protoc 4(4):470–483
21. Weynberg KD, Wood-Charlson EM, Suttle CA et al (2014) Generating viral metagenomes from the coral holobiont. Front Microbiol 5:206
22. Temmam S, Monteil-Bouchard S, Robert C et al (2015) Host-associated metagenomics: a guide to generating infectious RNA Viromes. PLoS One 10(10):e0139810
23. Froussard P (1993) rPCR: a powerful tool for random amplification of whole RNA sequences. PCR Methods Appl 2(3):185–190

Chapter 4

Phage on Tap: A Quick and Efficient Protocol for the Preparation of Bacteriophage Laboratory Stocks

Natasha Bonilla and Jeremy J. Barr

Abstract

A major limitation with traditional phage preparations is the variability in titer, salts, and bacterial contaminants between successive propagations. Here, we introduce the Phage On Tap (PoT) protocol for the quick and efficient preparation of homogenous bacteriophage (phage) stocks. This method produces homogenous, laboratory-scale, high titer (up to 10^{10-12} PFU/mL), endotoxin reduced phage banks that can be used to eliminate the variability between phage propagations, improve the molecular characterizations of phage, and may be applicable for therapeutic applications. The method consists of five major parts, including phage propagation, phage cleanup by 0.22 μm filtering and chloroform treatment, phage concentration by ultrafiltration, endotoxin removal, and the preparation and storage of phage banks for continuous laboratory use. From a starting liquid lysate of >100 mL, the PoT protocol generated a cleaned, homogenous, laboratory phage bank with a phage recovery efficiency of 85% within just 2 days. In contrast, the traditional method took upward of 5 days to produce a high titer, but lower volume phage stock with a recovery efficiency of only 4%. Phage banks can be further purified for the removal of bacterial endotoxins, reducing endotoxin concentrations by over 3000-fold while maintaining phage titer. The PoT protocol focused on T-like phages, but is broadly applicable to a variety of phages that can be propagated to sufficient titer, producing homogenous, high titer phage banks that are applicable for molecular and cellular assays.

Key words Bacteriophage, Endotoxin, Cesium chloride, Top agar, Ultrafiltration, Dialysis, Speed vacuum, Phage bank

1 Introduction

Due to increasing interest for the use of bacteriophage (phage) in medical, industrial, and molecular settings, new approaches are required to quickly and efficiently produce high titer, homogenous, and purified phage stocks. It is desirable that these stocks be free of bacteria, molds, debris, culture medium, and bacterial endotoxins [1]. Typically, the ability to produce high titer phage stocks is largely dependent on the particular phage and host cell under consideration, yet certain principles and methodologies can be broadly applied. Traditional techniques used for the concentration

Andrés Moya and Vicente Pérez Brocal (eds.), *The Human Virome: Methods and Protocols*, Methods in Molecular Biology, vol. 1838, https://doi.org/10.1007/978-1-4939-8682-8_4, © Springer Science+Business Media, LLC, part of Springer Nature 2018

and purification of phage involve centrifugation, filtration, ultrafiltration, precipitation with Polyethylene Glycol (PEG), ultracentrifugation in cesium chloride (CsCl) gradients, and dialysis [1–6]. Yet many of these techniques are time consuming and affect phage recovery and/or viability.

Phage preparations are often contaminated by macromolecules derived from the host bacteria and culture media, with the major pyrogen being the lipid A moiety of lipopolysaccharide (endotoxin) from the outer membrane of Gram-negative bacteria [7]. Endotoxins are amphipathic molecules; the lipid component is linked to a core polysaccharide and as a result they can form large aggregates greater than 1000 kDa in solution [8]. Endotoxin elicits a wide variety of pathophysiological effects in the body. Exposure to even small amounts can result in toxic shock, cell injury, cytokine production, and the activation of immune responses [9–12]. Due to these effects it is important that endotoxins be removed from phage preparations when studying or applying phage in the context of eukaryotic systems. The amount of endotoxin is defined as an endotoxin unit (EU), which corresponds to the activity of 100 pg of *E. coli* lipopolysaccharide. The endotoxin content of distilled water is estimated at 20 EU/mL, with the allowed limit for intravenous and oral administration set at 5 $EU{\bullet}kg{\bullet}h^{-1}$ and < 20 EU/mL respectively [13–15].

A further limitation with traditional phage preparations is the variability of titer, salts, and bacterial contaminants produced between successive propagations. Eliminating this variability has proven critical for accurate analysis for the molecular interactions of phage within the context of eukaryotic hosts [16, 17]. Here, we present the Phage On Tap (PoT) protocol as a fast and efficient way to produce homogenous laboratory phage stocks. Phage stocks were sterilized by centrifugation, 0.22 μm filtration, and chloroform treatment, before concentration and washing using ultrafiltration, and storage at 4 °C for 4 months with minimal degradation [18]. However, the Phage On Tap protocol was not able to effectively reduce bacterial endotoxins, likely due to our ultrafiltration-based approach and the large aggregate size of endotoxins. Numerous endotoxin removal procedures and commercial kits are available [8, 19–22], yet many of these methodologies lack generality, are time consuming, or are cost-prohibitive. Recently, the successful reduction of endotoxins (<20 EU/mL) from phage lysates was reported by extraction with organic solvents [23]. We corroborate this organic solvent-based method for the removal of bacterial endotoxin from phage lysates and adapt this methodology, with reduced processing time through the use of speed vacuum, in the Phage On Tap protocol (*see* **Note 1**). The PoT protocol can purify phage lysates, with volumes ranging between 50 and 300 mL (or greater if required), producing homogeneous phage stocks for further laboratory testing. The method takes 2 days to purify and concentrate a final phage lysate of ~10 mL with a titer >1010 PFU/mL, and a significant reduction in bacterial endotoxin levels [18]. Our method has been tested on a variety of phages, including T4 (*E. coli*), T3 (*E. coli*), T5 (*E. coli*), and Spp1 (*B. subtilis*), and is broadly applicable to other tailed phages. Recently, the method was used to clean and remove endotoxins from a therapeutic phage cocktail, which was subsequently used to successfully treat a patient with a disseminated multi-drug resistant *Acinetobacter baumannii* infection [24].

2 Materials

2.1 Equipment

1. 37 °C incubator with a rocker.
2. Centrifuge with swinging bucket rotor.
3. Microtube centrifuge.
4. Centrifugal vacuum concentrator.
5. Stir plate.
6. Petri dish with disposable lid.
7. 1.7 mL microcentrifuge tubes.
8. Falcon 50 mL conical centrifuge tubes.
9. Small glass test tubes 13 × 100 mm.
10. Serological pipettes.
11. 0.22 μm Sterivex filter units.
12. Whatman Anotop 0.02 μm sterile syringe filters.
13. Amicon® Ultra-15 centrifugal filter units, Ultracel 100 kDa membrane (*see* **Note 2**).
14. 500 mL PYREX® screw cap storage bottle with plastic seal.
15. Spectra-Por® Float-A-Lyzer® G2 Dialysis membrane, 10 mL, MWCO 3.5–5 kDa.
16. Nalgene™ General Long-Term Storage Cryo tubes.

2.2 Reagent Setup

1. LB broth: 25 g LB broth in 1 L of dH_2O.
2. LB top agar: 25 g LB broth, 7.5 g of agar, in 1 L of dH_2O.
3. LB agar plates: 25 g LB broth, 15 g of agar, in 1 L of dH_2O.
4. SM buffer: 5.8 g of NaCl, 2.0 g of $MgSO_4.7H_2O$, 50 mL of 1 M Tris–HCl pH 7.4, in 1 L of dH_2O. Autoclave, 0.02 μm filter-sterilize before use, and store at room temperature.
5. LB broth supplemented with $CaCl_2$ and $MgCl_2$: 100 mL of LB broth, 0.001 M $CaCl_2$, 0.001 M $MgCl_2$. Add in a 250 mL PYREX® screw cap storage, autoclave and store at room temperature.

3 Methods

Figure 1 illustrates the overall process for the Phage On Tap (PoT) phage purification protocol (*see* **Notes 3–5**).

3.1 Phage Plaque Assay for Determination of Titer [1]

1. Grow *E. coli* B bacterial host in LB broth overnight at 37 °C.
2. Dilute phage stock or isolate in LB broth down to the desired dilution (e.g., for a phage stock of 10^8 PFU/mL serially dilute down to 10^{-6} and 10^{-7} to obtain countable plaques).
3. Heat LB top agar in microwave until completely molten, then allow top agar to cool in a 56 °C water bath or until it is warm to the touch.
4. Add 1 mL of the overnight bacterial host and 1 mL of the phage dilution to a glass test tube and mix.
5. Add 3 mL of molten top agar to the glass test tube.
6. Quickly pour molten mixture onto an LB agar plate and tilt the plate to evenly distribute the agar. Let it sit undisturbed until the agar has gelled.

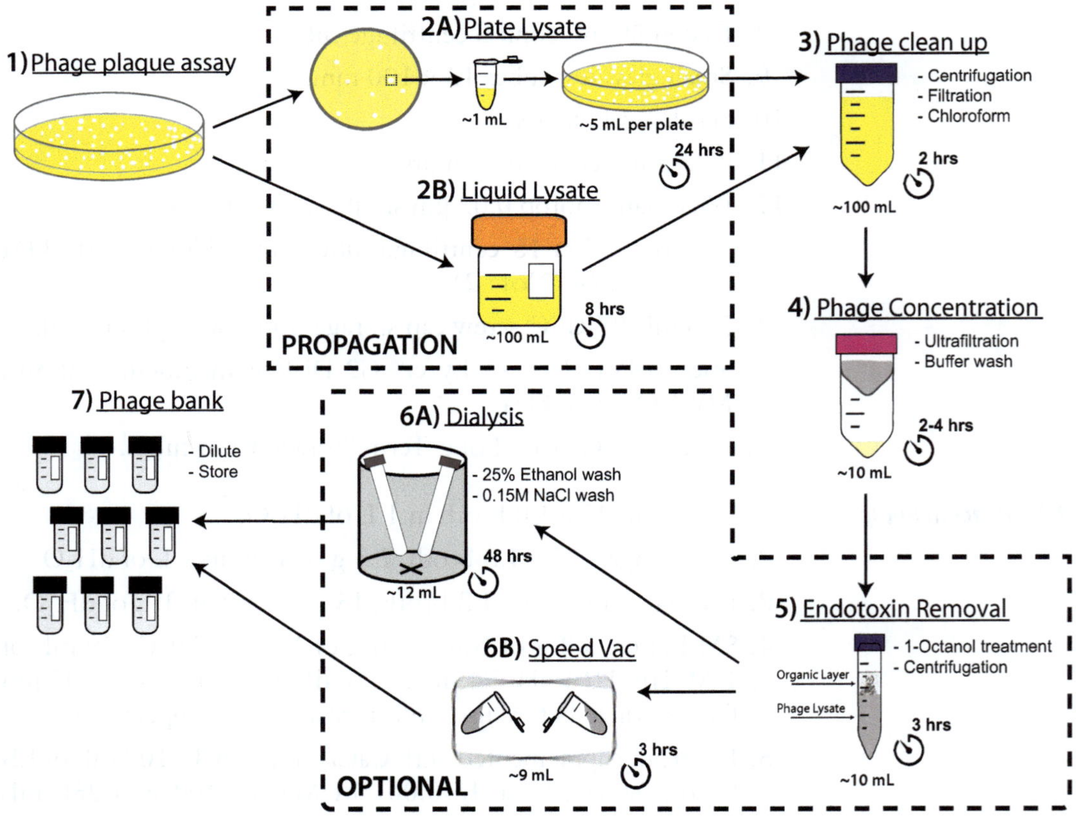

Fig. 1 Flow chart of the Phage on Tap (PoT) protocol for the production of high titer, homogenous, endotoxin reduced phage banks within 2 days. Phage titers (PFU/mL) and volumes (mL) were recorded after each step

7. Once the plate has gelled (~5 min), invert and incubate overnight at 37 °C.
8. Count phage plaques and determine phage titer in plaque-forming units (PFU/mL) using the following formula:

$$\text{PFU per mL} = \text{plaques per plate} \times \text{volume plated in mL} \times \text{dilution factor.}$$

e.g., if there are 20 plaques when you plated out 1 mL from the 10^6 dilution, the titer of the phage stock is 2.0×10^7 PFU/mL.

3.1.1 Phage Isolation and Propagation Via Plate Lysate

1. From a plate lysate plate pick a single phage plaque using a sterile Pasteur pipet.
2. Resuspend the plaque into a microcentrifuge tube containing 1 mL of filter-sterilized phage diluent (SM buffer) and vortex for 5 min.
3. Centrifuge at 4000 × *g* for 5 min to remove any remaining debris.
4. Perform plate lysate as described above using the resuspended phage.
5. After incubation the entire plate should be lysed. Pour ~5 mL of SM buffer on the top of the plate and shake gently for 15 min at room temperature.
6. Collect buffer from the top of the plate and centrifuge at 4000 × *g* for 5 min.
7. Collect phage lysate and store at 4 °C until cleanup.

Optional: Titer the lysate via plaque assay to ensure initial high titer.

3.1.2 Phage Propagation Via Liquid Lysate

1. Grow *E. coli* bacterial host in LB broth overnight at 37 °C.
2. Spike LB broth supplemented with $CaCl_2$ and $MgCl_2$ with 0.1 volumes of overnight bacterial host (*see* **Note 6**).
3. Incubate with agitation for 1 h at 37 °C.
4. Add 100 μL of high titer phage lysate (>10^8 PFU/mL).
5. Incubate at 37 °C with agitation for ~5 h or until lysate clears.
6. Collect the phage lysate and store at 4 °C until cleanup.

Optional: Titer the lysate via plaque assay to ensure initial high titer.

3.2 Phage Cleanup (0.22 μm Filtering and Chloroform)

1. Aliquot phage lysate into 50 mL sterile falcon centrifuge tubes and centrifuge at 4000 × *g* for 20 min.
2. Carefully collect the supernatant using a serological pipette and transfer into a properly labeled sterile falcon tube.

3. Filter-sterilize the phage supernatant using a 0.22 μm filter to yield a bacterial cell-free phage lysate.
4. Add 0.1 volumes of chloroform to the supernatant, vortex, and incubate at room temperature for 10 min (*see* **Note 7**). CAUTION: Chloroform is toxic and should only be used in a fume hood and with personal safety equipment, such as gloves and goggles.
5. Centrifuge at 4000 × *g* for 5 min and transfer the supernatant into 250 mL PYREX® screw cap storage bottle and store at 4 °C until concentration.

Optional: Titer the lysate via plaque assay to ensure initial high titer.

3.3 Phage Concentration and Wash Via Ultrafiltration

1. Add ~15 mL of phage lysate into the upper reservoir of Amicon filtration device.
2. Centrifuge Amicon at 4000 × *g* for ~5 min (*see* **Note 8**).
3. Carefully discard the filtrate into a waste bucket and add another volume of phage lysate to the sample filter cup and repeat centrifugation (*see* **Note 9**).
4. Repeat step until all phage lysate has been concentrated to <10 mL.
5. Add ~15 mL of SM buffer into the upper reservoir containing concentrate phage lysate and centrifuge at 4000 × *g* for ~5 min to wash phage lysate (*see* **Note 10**).
6. Repeat wash step and concentrate washed phage lysate to <10 mL.
7. Using a pipette, carefully collect phage lysate from the upper reservoir and gently wash the surface of the upper reservoir.
8. Collect <10 mL of concentrated and purified phage lysate. Titer phage concentrate and record PFU/mL.

3.4 Endotoxin Removal [23, 25] (*See* Notes 11–13)

1. Add 0.4 volumes of 1-octanol to phage concentrate and shake at room temperature for 1 h.
2. Incubate phage concentrate at 4 °C for 1.5 h.
3. Centrifuge at 4000 × *g* for 10 min.
4. Using a syringe, pierce the bottom of the tube and collect the aqueous phase (bottom layer) that contains your phage, and transfer to a sterile 50 mL falcon tube. Do not collect the organic phage (top layer) or interface as this contains endotoxins (*see* **Notes 14** and **15**).

3.5 Dialysis Removal of Organic Solvent [23] (*See* Notes 16 and 17)

1. Pre-wet Spectra-Por® dialysis membrane with sterilized dH_2O according to the manufacturer's instructions, being careful not to contaminate the inside of the tubing.

2. Load maximum of 10 mL of phage concentrate inside of dialysis tubing and seal tightly.
3. Dialyze phage concentrate against 2 L of 25% (v/v) ethanol at 4 °C on a stirring plate for 24 h to remove residual 1-butanol, replacing the 25% ethanol solution four times at 15 h, 18 h, 21 h, and 24 h.
4. Dialyze phage concentrate against 2 L of 0.15 M NaCl solution at 4 °C on a stirring plate 24 h to remove residual ethanol, replacing the 0.15 M NaCl solution three times after 15 h, 19 h, and 24 h.
5. Carefully collect the phage concentrate by washing the inside of the dialysis tubing. Store concentrate at 4 °C until phage bank preparation.

Optional: Titer the lysate via *plate lysates to ensure high titer. Perform the EndoZyme® Recombinant Factor C Endotoxin Detection Assay according to the manufacturer's instructions to obtain quantitative endotoxin levels.*

3.6 Speed Vacuum Removal of Organic Solvent (See Note 18)

1. Aliquot ~1 mL of the phage concentrate equally into microcentrifuge tubes.
2. Place tubes into a speed vacuum 4 °C, open lids, and centrifuge at 4000 × *g* for 3 h.
3. After speed vacuum, phage concentrate volume should be reduced by approximately 30% and residual 1-octanol evaporated.
4. Collect the phage concentrate and store 4 °C until phage bank preparation.

Optional: Titer the lysate via *plaque assay to ensure high titer. Perform the EndoZyme® Recombinant Factor C Endotoxin Detection Assay according to the manufacturer's instructions to obtain quantitative endotoxin levels from phage concentrates.*

3.7 Phage Bank Storage

1. Dilute phage lysate in SM buffer to generate a high-titer working stock (*see* **Note 19**).
2. Titer diluted phage concentrate and record PFU/mL.
3. Aliquot phage working stocks into labeled cryo tubes and store at 4 °C.
   - You now have a bank of ho.mogenous, high titer (up to 10^{10-11} PFU/mL) phage bank for laboratory testing.
   - From our experience the use of homogenous and standardized phage banks can significantly reduce the experimental variability for the molecular characterization of phages.

4 Notes

1. Endotoxin Quantitation Kit is optional and is only required for the quantitation of endotoxins in phage lysates.
2. 100 kDa membrane pore size is equivalent to a spherical particle with a diameter of ~3 nm and is therefore sufficient for the capture of all known bacteriophages[26].
3. The procedure for phage propagation is largely specific for each phage and bacterial host. Here, we use propagation conditions for T4 phage and *Escherichia coli* B bacterial host. It is recommended to use appropriated growth and propagation conditions for your choice of phage and host.
4. Once a sufficiently high titer phage lysate is obtained please proceed to Subheading 3, **step 2**.
5. It is recommended to only propagate and purify one phage at a time to prevent cross-contamination.
6. For a phage with a high burst size, such as T3, you may need to double volume of host added.
7. Lipid enveloped phages are sensitive to chloroform and titer can be significantly reduced. If a drop in titer is observed skip this step.
8. Centrifugation times will vary based on phage type and titer. It is important not to spin the device dry. If unsure about centrifugation times select a shorter spin time, check the lysate level, and adjust spin times appropriately.
9. The same device can be used to concentrate large volumes of phage lysate (>100 mL) achieving an approximate 90% decrease with minimal loss in phage titer. If the device clogs or concentration significantly slows, transfer phage supernatant to a new device and continue.
10. SM buffer was chosen as it is suitable for the long-term storage of T-phage, but any appropriate storage buffer can be used. It is important not to spin the device dry, adjust centrifugation times accordingly.
11. This method is adapted from Szermer-Olearnik and Boratyński [23], which demonstrates the efficient removal of endotoxins from bacteriophage lysates using water immiscible solvents that are subsequently removed via dialysis. For detailed explanation of the methodology please *see* Szermer-Olearnik and Boratyński [23] and Morrison and Leive [25].
12. Our adapted method uses a speed vacuum to remove residual organic solvent from phage lysates, instead of the lengthy dialysis washes with similar efficiency.

13. This step is optional. If you do not require removal of bacterial endotoxins from your phage preparations, please go to **step 6** in Subheading 3.
14. It is best to leave a small residual amount of phage concentrate behind to reduce the transfer of contaminating endotoxins.
15. If endotoxin removal is not sufficient the method can be repeated to further reduce endotoxins.
16. This method is adapted from Szermer-Olearnik and Boratyński [23] and describes the removal of residual organic solvents from phage lysates by dialysis.
17. Due to the ionic concentration of phage SM buffer used you may end up with greater than the starting volume.
18. This method is a faster alternative to the dialysis method for the removal of residual organic solvents from phage concentrates.
19. This step largely depends on desired concentration and volume of the phage bank. If you require higher titer phage stocks then dilute less or omit dilution, if greater volume is desired then dilute more.

References

1. Adams MH (1959) Bacteriophages. Interscience, New York, NY
2. Yamamoto KR, Alberts BM, Benzinger R et al (1970) Rapid bacteriophage sedimentation in the presence of polyethylene glycol and its application to large-scale virus purification. Virology 40:734–744
3. Seeley ND, Primrose SB (1982) A review: the isolation of bacteriophages from the environment. J Appl Bacteriol 53:1–17
4. Suttle CA, Chan AM, Cottrell MT (1991) Use of ultrafiltration to isolate viruses from seawater which are pathogens of marine phytoplankton. Appl Environ Microbiol 57:721–726
5. Carlson K (2004) Working with bacteriophages: common techniques and methodological approaches. CRC Press
6. Bourdin G, Schmitt B, Marvin Guy L et al (2014) Amplification and purification of T4-like *Escherichia coli* phages for phage therapy: from laboratory to pilot scale. Applied Environ Microbiol 80:1469–1476
7. Raetz CRH, Reynolds CM, Trent MS et al (2007) Lipid a modification systems in gram-negative bacteria. Annu Rev Biochem 76:295–329
8. Magalhães PO, Lopes AM, Mazzola PG et al (2013) Methods of endotoxin removal from biological preparations : a review. J Pharm Pharm Sci 10:388–404
9. Morrison DC, Ulevitch RJ (1978) The effects of bacterial endotoxins on host mediation systems. A review. Am J Pathol 93:526–618
10. Rietschel E, Kirikae T, Schade F et al (1994) Bacterial endotoxin: molecular relationships of structure to activity and function. FASEB J 8:217–225
11. Alexander C, Rietschel ET (2001) Invited review: bacterial lipopolysaccharides and innate immunity. J Endotoxin Res 7:167–202
12. Pabst MJ, Pabst KM, Handsman DB et al (2008) Proteome of monocyte priming by lipopolysaccharide, including changes in interleukin-1beta and leukocyte elastase inhibitor. Proteome Sci 6:13
13. Bruttin A, Brüssow H (2005) Human volunteers receiving *Escherichia coli* phage T4 orally: a safety test of phage therapy. Antimicrob Agents Chemother 49:2874–2878
14. Gorbet MB, Sefton MV (2005) Endotoxin: the uninvited guest. Biomaterials 26:6811–6817
15. Abedon ST, Kuhl SJ, Blasdel BG et al (2011) Phage treatment of human infections. Bacteriophage 1:66–85

16. Barr JJ, Auro R, Furlan M et al (2013) Bacteriophage adhering to mucus provide a non-host-derived immunity. Proc Natl Acad Sci U S A 110:10771–10776
17. Barr JJ, Auro R, Sam-Soon N et al (2015) Subdiffusive motion of bacteriophage in mucosal surfaces increases the frequency of bacterial encounters. Proc Natl Acad Sci U S A 112:201508355
18. Bonilla N, Rojas MI, Netto Flores Cruz G et al (2016) Phage on tap–a quick and efficient protocol for the preparation of bacteriophage laboratory stocks. PeerJ 4:e2261
19. Boratyński J, Syper D, Weber-Dabrowska B et al (2004) Preparation of endotoxin-free bacteriophages. Cell Mol Biol Lett 9:253–259
20. Merabishvili M, Pirnay J-P, Verbeken G et al (2009) Quality-controlled small-scale production of a well-defined bacteriophage cocktail for use in human clinical trials. PLoS One 4:e4944
21. Oślizło A, Miernikiewicz P, Piotrowicz A et al (2011) Purification of phage display-modified bacteriophage T4 by affinity chromatography. BMC Biotechnol 11:59
22. Branston SD, Wright J, Keshavarz-Moore E (2015) A non-chromatographic method for the removal of endotoxins from bacteriophages. Biotechnol Bioeng 112:1714–1719
23. Szermer-Olearnik B, Boratyński J (2015) Removal of endotoxins from bacteriophage preparations by extraction with organic solvents. PLoS One 10:e0122672
24. Schooley RT, Biswas B, Gill JJ et al (2017) Development and use of personalized bacteriophage-based therapeutic cocktails to treat a patient with a disseminated resistant *Acinetobacter baumannii* infection. Antimicrob Agents Chemother AAC.00954–17
25. Morrison DC, Leive L (1975) Fractions of lipopolysaccharide from *Escherichia coli* O111:B4 prepared by two extraction procedures. J Biol Chem 250:2911–2919
26. Erickson HP (2009) Size and shape of protein molecules at the nanometer level determined by sedimentation, gel filtration, and electron microscopy. Biol Proc Online 11:32–51

Part II

Gastrointestinal Tract

Chapter 5

Extraction and Purification of Viruses from Fecal Samples for Metagenome and Morphology Analyses

Josué L. Castro-Mejía, Ling Deng, Finn K. Vogensen, Alejandro Reyes, and Dennis S. Nielsen

Abstract

The human enteric virome consists of endogenous retro elements and viruses that infect the host and members of the gut microbiome (GM). Mounting evidence suggests that the gut virome plays a central role in maintaining homeostasis and via the GM influences immunology of the host. To thoroughly characterize the gut virome, it is often very useful to first separate and concentrate extracellular viral-like particles (eVLPs) enabling an integrative characterization of them. Here, we describe a detailed protocol for extraction and concentration of the viral fraction from fecal samples based on a polyethylene glycol precipitation (PEG) approach. These procedures maximize the yields of eVLPs (and their DNA) with high purity well suited for down-stream analysis such as quantification and morphological assessment, determination of phage-host pairs as well as virome sequencing.

Key words Gut-virome, Extraction procedures, Metagenome of extracellular viral-like particles (eVLPs), Morphology, Polyethylene-glycol (PEG) precipitation

1 Introduction

During the last decade, it has become evident, that gut microbiota (GM) dysbiosis is associated with a range of diseases [1]. Further, during the last couple of years, it has become recognised, that dysbiosis of the prokaryotic component of the GM is often linked with alterations in the composition of gut viral communities (also termed "virome"). Gut virome alterations have been linked to the etiology of severe acute malnutrition (SAM), irritable bowel diseases (IBD), and type-1 diabetes (T1D). The gut virome is highly dynamic within the first couple of years after birth, probably playing a role in early life GM development, as well as in the protection against harmful bacteria via the so-called "bacteriophage adhering to mucus" (BAM) immunity [2–6].

Still, unlike other components of the human gut microbiome (GM), the gut virome remains relatively poorly characterized.

Andrés Moya and Vicente Pérez Brocal (eds.), *The Human Virome: Methods and Protocols*, Methods in Molecular Biology, vol. 1838, https://doi.org/10.1007/978-1-4939-8682-8_5,

Viruses lack of a universal gene-marker [7], furthermore, viruses are often agents of genetic transfer thus carrying host genetic material and, in particular phages, are usually present in lysogenic states integrated within the bacterial genome. It is known that up to 10% of the bacterial genomes may be composed of viral material, thus raising challenges in the differentiation of viral material from bacterial/host signals in metagenomic datasets; not forgetting the limited annotation of viral genomes currently available in public databases. Further, a large fraction of the viruses in the human gut remains to be described and annotated which hampers data mining and intepretation [8]. Such aspects indicate the strict need of applying metagenome analysis on extracellular viral-like particles (eVLPs), as well as the implementation of standardized procedures that ensure high levels of virions recovery and purity in an unbiased manner.

The basis for concentration and extraction of eVLPs and detailed workflows for sample preparation has been thoroughly presented elsewhere [9–14]. Despite the fact that those approaches may be fast and straightforward they are not fully adapted to deal with complex matrices like intestinal material. Viruses differ in size and structure (*see* Fig. 1), in consequence extraction procedures have been shown to influence (a) the yields of eVLPs and associated DNA, (b) the probability of retaining rare viruses as well as, (c) the purity of the extraction. All these are important components for metagenome and morphological analyses [15] and may arguably represent the most critical aspects of any viral metagenome approach on any complex microbial assemblage [16].

Here, we describe a detailed protocol for extraction and concentration of eVLPs from fecal samples based on the polyethylene glycol precipitation (PEG) approach. This protocol was recently optimized [15] and compared to control procedures [9] with the optimized protocol yielding 16 times more phage particles and up to 68 times more viral DNA. eVLPs extracted through this protocol can successfully be subjected to transmission electron microscopy (TEM), epifluorescence microscopy (EM), eVLPs-propagation (for known hosts or host-searching), and diverse molecular genetic analysis including virome sequencing.

2 Materials

All solutions must be prepared with ultra-pure water (holding a resistivity of 18.2 MΩ·cm at 25 °C) using only analytical grade reagents. Prepare and store reagents at room temperature. As part of preparation, buffer solutions as well as other (re-usable) materials must be autoclaved (121 °C for 15 min).

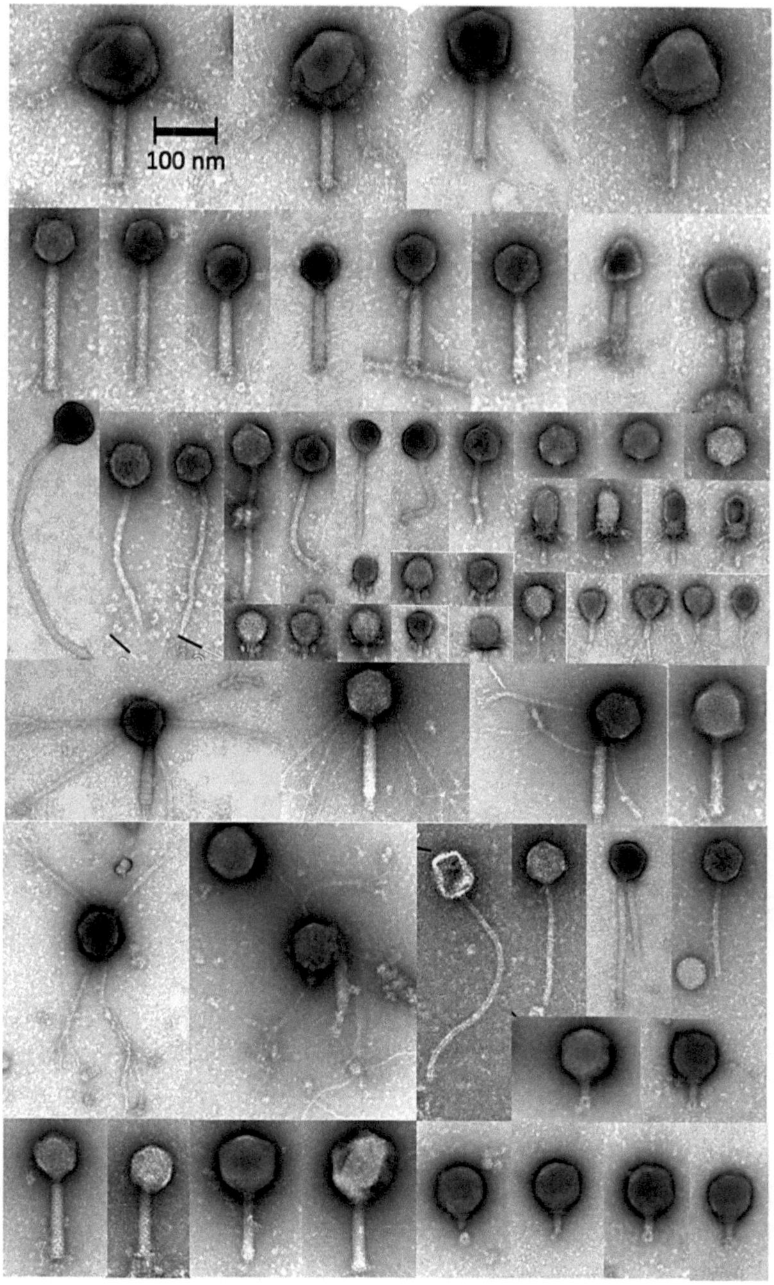

Fig. 1 Electron micrographs demonstrate the great diversity (size and structure) of the human gut virome. The displayed morphotypes are members of *Myoviridae*, *Siphoviridae*, and *Podoviridae* viral families, as well as putative unclassified ones. (Source: Castro-Mejía et al. [15])

1. SM Buffer (per 1 L of solution): 50 mM Tris–HCl, pH 7.5, 200 mM NaCl, 10 mM $MgSO_4$. Autoclave 121 °C, 15 min.
2. Cesium Chloride (CsCl) in SM buffer (per 100 mL of solution) with different densities g/cm^3 (ρ).

(a) 100 mL of SM buffer.
(b) Ultra-Pure CsCl (optical-grade).
- For 1.39ρ, 52.86 g CsCl in 100 mL solution (refractive index, 1.371).
- For 1.70ρ, 95.14 g CsCl in 100 mL solution (refractive index, 1.399).

Autoclave 121 °C, 15 min.

3. *See* **Note 1** for detailed preparation.

2.1 Specialist Equipment

1. Seward-Stomacher (or similar mixer) and filter-bags (e.g., Interscience® micro-perforated filter).
2. Ultracentrifuge.
3. Refractometer.

2.2 Other Materials

1. Glass fiber filters.
2. Polyethylene glycol (PEG) 6000.
3. 100 kDa Dialysis membranes.
4. 50 mL Falcon tubes or similar.
5. 17 mL tubes for ultracentrifugation.
6. 23 cm glass Pasteur pipettes (they need to be sterilized).
7. 5 mL syringe and needles (ø 0.8 mm).

3 Methods

Perform all steps at room temperature unless otherwise stated. The basis of the protocol relies on the extraction from 5 g of fecal material, but proportional adaptations can also be applied. We have applied the current method to a minimum of 50 mg with consistent results.

3.1 Pre-processing of Samples

1. Five grams of fecal material are weighted in a filter-bag, add 45 mL of SM buffer (90% v/w) and mix in a Seward-Stomacher for 2 min at medium speed.
2. The filtered fractions (~45–50 mL) are transferred to a 50 mL Falcon tube and centrifuge at 5000 × *g* for 45 min at 4 °C. Carefully, the supernatant (~38–45 mL) is collected and placed it into a new tube (do NOT disturb any fraction of the large pellet, *see* **Note 2**).
3. The supernatant is passed through a glass fiber filter membrane (~34–38 mL are collected).

3.2 Purification of Samples

1. The pre-processed sample is centrifuged at 5000 × *g* for 45 min at 4 °C, carefully collect the supernatant (~30–33 mL), and place it into a new tube (do NOT disturb any fraction of the large pellet! *see* **Note 2**).
2. Subsequently, the sample is supplemented with 10% (v/w) of PEG 6000 (*see* **Note 3**), mixed gently (until dissolution), and kept at 4 °C for at least 2 h.
3. Afterward the sample is centrifuged at 25,000 × *g* for 45 min at 4 °C. The supernatant is discarded (*see* **Note 4**) and the pellet resuspended (*see* **Note 5**) and dissolved overnight in 7 mL of SM buffer.
4. The resuspended pellet is then dialyzed (100 kDa) in 350 mL SM buffer overnight. Exchange SM buffer once after 2 h of dialysis).
5. After dialysis, the sample's volume is adjusted to 10 mL, centrifuged at 5000 × *g* for 45 min at 4 °C, and the supernatant (~7–8 mL) is carefully retrieved and placed into a new tube (do not disturb any fraction of the large pellet! *see* **Note 6**).
6. To prepare a two-layer CsCl gradient, the volume of the sample is brought up to 14 mL and its density adjusted to 1.39ρ (*see* **Note 7**). Then, the adjusted sample is mounted on the top of 2 mL 1.7ρ CsCl solution into an ultracentrifugation tube (Fig. 2 and *see* **Note 8**) and centrifuged at 150,000 × *g* for 4.5 h at 4 °C.
7. The density fraction between 1.41 and 1.52ρ is recovered (with syringe and needle, *see* Fig. 3) and subsequently placed in sterile containers and stored at 4 °C until further use (e.g., EM (*see* Fig. 4)), TEM (Fig. 1) and extraction of DNA prior DNase treatment.

4 Notes

1. It is recommended to place the weighted CsCl in a volumetric flask and bring the volume up to 100 mL, to reach an appropriated density and/or refractive index.
2. The supernatant must be collected with 5–10 mL tips in order to prevent decanting of the centrifuged sample that may disturb the pellet. Disruption of any fraction of the pellet must be avoided as it contains large amount of impurities and contaminant cells. In case of disruption, the sample must be centrifuged for another 10–15 min.
3. PEG 6000 should be dissolved by gently mixing. Violent mixing can be detrimental for the integrity of the viral capsids and leads to the release of viral DNA.

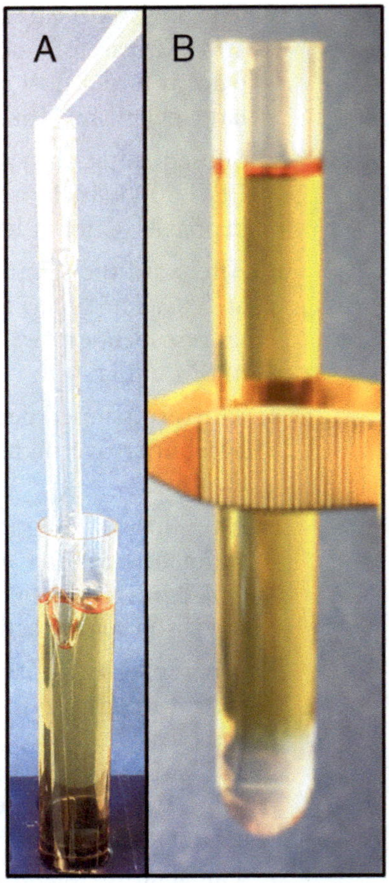

Fig. 2 Method and setup for buoyant density ultracentrifugation. eVLPs are separated and concentrated using differential centrifugation based on gradient of CsCl. (**a**) 2 mL of 1.7ρ CsCl are placed in the bottom of the ultracentrifugation tube using a Pasteur pipette, (**b**) which remains in equilibrium (notice the density interface by changes in color) under 14 mL of 1.39ρ adjusted sample

4. The supernatant must be removed using 5–10 mL tips in order to prevent decanting the tube and losing the pellet containing the viral particles.
5. If possible, pellet-clump should be detached from the bottom of the tube and let it dissolve overnight without mixing.
6. Similar to **Note 2**, this step is crucial to avoid presence of contaminant cells in downstream steps.
7. Typically, the density of the sample is around 1.01–1.06ρ and can be brought up to 1.39ρ using CsCl in SM buffer previously adjusted to 1.70ρ.
8. To perform a two-layer gradient, it is recommended to use sterilized glass Pasteur pipettes in order to allocate 1.70ρ CsCl at the bottom of the tubes as shown in Fig. 2.

Fig. 3 After ultracentrifugation (150,000 × *g* for 45 min at 4 °C) the resulting gradient localizes the eVLPs between 1.41 and 1.52ρ. In this step, this density fraction needs to be check by measuring the refractive index in different sections of the gradient-column formed (typically 1.53ρ is located approximately 1 cm from the bottom of the 17 mL tube). A sterile needle is then inserted (pierced) into the tube at the level of 1.53ρ to collect (typically 1 mL) of the fraction of concentrated eVLPs

Acknowledgments

This work was supported by "CALM" – Counteracting Age-related Loss of Skeletal Muscle Mass (J.L.C.) and "EarlyVir—Influence of diet on early life gut virome—a key player in shaping the gut microbiota". Joint Programming Initiative, Healthy Diet for a Healthy Living program (L.D.)

Fig. 4 Epifluorescence Microscopy (EM) to determine eVLPs load and purity of fecal viromes isolated by the procedures herein described. EM analysis can be used as a quality-control step for checking the eVLPs purity, absence of contaminant cells and quantification of eVLPs. According to our observations the yields of eVLPs could vary from 10^8–10^{10} eVLPs per gram of fecal matter

References

1. Wu H, Tremaroli V, Bäckhed F (2015) Linking microbiota to human diseases : a systems biology perspective. Trends Endocrinol Metab 26:758–770
2. Reyes A, Blanton LV, Cao S et al (2015) Gut DNA viromes of Malawian twins discordant for severe acute malnutrition. Proc Natl Acad Sci U S A 112:1941–11946
3. Zhao G, Vatanen T, drit L et al (2017) Intestinal virome changes precede autoimmunity in type I diabetes-susceptible children. Proc Natl Acad Sci U S A 114:E6166–E6175
4. Barr JJ, Auro R, Furlan M et al (2013) Bacteriophage adhering to mucus provide a non-host-derived immunity. Proc Natl Acad Sci U S A 110:10771–10776
5. Norman JM, Handley SA, Baldridge MT et al (2015) Disease-specific alterations in the enteric virome in inflammatory bowel disease. Cell 160:447–460
6. Lim ES, Zhou Y, Zhao G et al (2015) Early life dynamics of the human gut virome and bacterial microbiome in infants. Nat Med 21:1228–1234
7. Sullivan MB (2015) Virome, not gene markers, for studying double-stranded DNA virus communities. J Virol 89:2459–2461
8. Hurwitz BL, Ren JMU, Youens-clark K (2016) Computational prospecting the great viral unknown. FEMS Microbiol Lett 363:fnw077
9. Thurber RV, Haynes M, Breitbart M et al (2009) Laboratory procedures to generate viral metagenomes. Nat Protoc 4:470–483
10. Wommack KE, Williamson, KE, Helton, RR et al (2009). Mehtods for the isolation of viruses from environmental samples. Methods Mol Biol 501:3–14
11. Conceição-Neto N, Zeller M, Lefrêre H et al (2015) Modular approach to customise sample preparation procedures for viral metagenomics: a reproducible protocol for virome analysis. Nat Publ Gr 5:16532
12. Hall RJ, Wang J, Todd AK et al (2014) Evaluation of rapid and simple techniques for the enrichment of viruses prior to metagenomic virus discovery. J Virol Methods 195:194–204

13. Kleiner M, Hooper LV, Duerkop BA (2015) Evaluation of methods to purify virus-like particles for metagenomic sequencing of intestinal viromes. BMC Genomics 16:7
14. Hoyles L, McCartney AL, Neve H et al (2014) Characterization of virus-like particles associated with the human faecal and caecal microbiota. Res Microbiol 165:803–812
15. Castro-Mejía J, Muhammed MK, Kot W et al (2015) Optimizing protocols for extraction of bacteriophages prior to metagenomic analyses of phage communities in the human gut. Microbiome 3:64
16. Hayes S, Mahony J, Nauta A et al (2017) Metagenomic approaches to assess bacteriophages in various environmental niches. Viruses 9:127

Chapter 6

Virome Sequencing of Stool Samples

Lenka Kramná and Ondřej Cinek

Abstract

Next-generation sequencing has opened avenues to studying complex populations such as the bacteriome (all bacteria), mycobiome (all fungi), and virome (all viruses in a given sample). Viromes are less often investigated as compared to bacteriomes. The reasons are mostly methodological: because no common pan-viral sequence signature exists, metagenomic sequencing remains the only option. This brings about the need of laborious virus enrichment, multiple signal amplification steps with virtually no possibility of interim quality control, and complicated bioinformatic analysis of the ensuing sequence data. Nevertheless, over the past decade virome sequencing has been enormously successful in identifying new agents in human and animal diseases, and in characterizing viruses in various ecological niches. Recently, virome sequencing has been also employed in studies of non-infectious diseases, which has brought about new challenges of sensitivity, costs, and reproducibility in testing of large sets of samples. Here, we present a detailed protocol that has been utilized in virome studies where hundreds of samples had to be reliably tested in order to assess the association of the stool virome with susceptibility to type 1 diabetes, a non-infectious autoimmune disease.

Key words Metagenome, Virus, Bacteriophage, Stool

1 Introduction

Metagenomic shotgun sequencing is a demanding laboratory and analytical technique. In virome research it is further complicated by the varying size of virus genomes, their high mutation rate, and the limited knowledge of viral genomes. The virus concentration in a sample is usually insufficient for existing commercially available next-generation library preparation kits, and therefore unbiased amplification steps must be employed. Furthermore, in a situation when many samples are sequenced together, the presently available sample indexing strategies show certain limitations in distinguishing the origin of signal in situations when samples with billions of virus copies are sequenced along with negative ones.

This protocol presents a procedure of virome sequencing from stool samples. It uses mechanical virus enrichment, random fragment formation, and PCR-based amplification. The procedure was

Andrés Moya and Vicente Pérez Brocal (eds.), *The Human Virome: Methods and Protocols*, Methods in Molecular Biology, vol. 1838, https://doi.org/10.1007/978-1-4939-8682-8_6,

adopted from studies of infectious diseases and viral ecology [1–9] and modified for improved performance in larger sample sets where reproducibility is needed for association studies of virome components with susceptibility to non-infectious diseases [10].

The protocol starts with simple mechanical steps enriching the virus fraction in the sample, and continues with nucleic acid extraction, sequence-independent single primer amplification, indexing, and library equalization. Critical points of the procedure are commented in notes.

2 Materials

2.1 Instruments

2.1.1 Pre-amplification Area

1. Thermocycler.
2. Heat-block for 1.5 mL tubes.
3. Refrigerated centrifuge for 1.5 mL tubes (10,000 × *g*).
4. Refrigerated centrifuge for PCR plates.
5. Ultracentrifuge or high-speed centrifuge (>50,000 × *g*) with tubes (at least 1 mL) (*see* **Note 1**).

2.1.2 Post-amplification Area

1. Thermocycler.
2. Real-time thermocycler.
3. Qubit fluorometer with tubes (Qubit assay tubes).
4. Agilent 2100 Bioanalyzer.
5. Equipment for agarose gel electrophoresis.
6. DynaMag-96 Side Magnet (*or* optionally, a Biomek 3000/4000 series robot).

2.2 Chemicals and Consumables

2.2.1 Virus Enrichment, Purification, and Nucleic Acid Extraction

1. HBSS (Hank's Balanced Salt Solution).
2. RNAse cocktail.
3. DNase Turbo.
4. Benzonase.
5. QIAamp Viral RNA Mini Kit.
6. QuantiTect Probe RT-PCR Kit (optional).
7. Armored RNA Enterovirus (optional).
8. Corning Costar Spin-X centrifuge tube filters, cellulose acetate membrane, sterile, pore size 0.45 and 0.22 μm.

2.2.2 Random Fragments Preparation and Amplification

1. Superscript III Reverse Transcriptase.
2. Recombinant RNasin Ribonuclease Inhibitor.
3. 10 mM dNTP.
4. Klenow Fragment, Exonuclease Minus.

5. SAP (shrimp alkaline phosphatase).
6. Exonuclease I.
7. AmpliTaq Gold DNA Polymerase.
8. *Primers* (*see* **Note 2**): D2_8N (AAGCTAAGACGGCG GTTCGGNNNNNNNN) and D2 (AAGCTAAGACGGCG GTTCGG).

2.2.3 Purification and Dilution Before Library Construction

1. Agencourt AMPure.
2. Quant-iT™ dsDNA High-Sensitivity Assay Kit.
3. 75% ethanol (freshly prepared, 360 μL per sample, plus excess).

2.2.4 Library Construction Using the Nextera XT Kit

1. Nextera XT DNA Sample Preparation Kit (24).
2. Nextera® XT Index Kit v2 Set A and/or Nextera® XT Index Kit v2 Set D.
3. TruSeq Index Plate Fixture (recommended).

2.2.5 Purification, Normalization, and Pooling of the Libraries

1. Agencourt AMPure.
2. 80% ethanol (freshly prepared, 360 μL per sample, plus excess).
3. KAPA library quantification kit.
4. Agilent High Sensitivity DNA Kit.
5. Deepwell plate 96/1000 μL, DNA LoBind).
6. Aluma Seal II or similar adhesive aluminum foil.

2.2.6 Denaturation and Sequencing of the Pooled Libraries

1. MiSeq reagent kit v2 2 × 250 bp.
2. PhiX control v3.
3. 10 M NaOH.
4. 35% HCl.

3 Methods

3.1 Virus Enrichment, Purification, and Nucleic Acid Extraction

Viruses and virus-like particles (*see* **Note 3**) are enriched, and the total nucleic acid is extracted. The input material for this section is a buffered 10% stool suspension. The virus enrichment starts with the removal of rough contaminants by several centrifugation steps, continues with filtering to remove bacterial cells, and is completed by pelleting of viral and virus-like particles by ultracentrifugation. The pellet is then resuspended. Before extraction of DNA and RNA, cocktail of nucleases can optionally be applied to digest unencapsulated free nucleic acid (i.e., genetic material not protected in viral capsids). The total RNA/DNA is then extracted.

3.1.1 Procedure

The centrifuge, ultracentrifuge, and all rotors should be prechilled. The maximum count of samples per batch is usually determined by the capacity of the ultracentrifuge. Systematic use of positive and negative controls in each run of the enrichment is encouraged, but probably unrealistic if only a low-capacity ultracentrifuge is available.

1. Let the samples thaw (*see* **Note 4**) and vortex them well (at least 15 s of pulse vortexing).
2. Centrifuge them in their original tubes, 10,000 × *g*, 2 min, 8 °C.
3. Transfer 800 μL supernatant into another tube.
4. Spin the supernatant similarly.
5. Transfer 750 μL supernatant into another tube.
6. Spin the supernatant similarly.
7. Transfer 700 μL supernatant onto a bacterial 0,45 μm spin filter.
8. Spin at 10,000 × *g*, 2 min, 8 °C (*see* **Note 5**).
9. Transfer the whole filtrate onto a 0.22 μm spin filter.
10. Spin at 3000 × *g*, 2 min, 8 °C.
11. Prefill ultracentrifugation tubes with 500 μL HBSS (*see* **Note 6**). Add 500 μL of the filtrate from the previous step and mix well by pipetting.
12. Perform ultracentrifugation at 750,000 × *g* (*see* **Note** 7), 1 h, at 8 °C, vacuum (*see* **Note 8**).
13. After ultracentrifugation, slowly decant or gently aspirate the supernatant. Do not disturb the pellet (*see* **Note 9**).
14. Add 500 μL of HBSS to the pellet and resuspend it by slow pipetting.
15. Transfer the whole volume into a 1.5 mL tube.
16. Here the procedure can be continued with:
    - immediate RNA/DNA extraction
    - storing the sample at −20 °C to −80 °C with deferred RNA/DNA extraction
    - digestion of unecapsulated nucleic acids immediately followed by RNA/DNA extraction.

    Optional: Digest unencapsulated nucleic acids (*see* **Note 10**):

    - Prepare a mix of nucleases for the total count of samples + excess. For each sample prepare 12 μL of 10× Turbo DNAse buffer, 2.4 μL of Turbo DNAse (2 U/μL), 8.0 μL of RNAse cocktail (500 U/mL RNase A + 20,000 U/mL Rnase T1), and 1.0 μL of Benzonase (5 U/μL).

- Combine 23.4 μL of this nuclease mix with 120 μL of the filtered sample.
- Mix by pipetting.
- Incubate at 37 °C, for 2 h.
- This incubation must be immediately (!) followed by nucleic acid extraction in order to stop the digestion.

17. Extract viral RNA/DNA with QIAamp Viral RNA Mini Kit, always without the use of carrier RNA (*see* **Note 11**) (the starting material is either the product of the digestion, or 140 μL filtered material). Store the extracted nucleic acid at −20 – −80 °C until further processing (**SAFE STOPPING POINT**).

    Optional: If positive controls were used in the process, quantify them now.

3.2 Random Fragments Preparation and Amplification

Random fragments of total nucleic acid are prepared, amplified and subsequently selected by size. The input material for this section is the total RNA/DNA of the virus-enriched sample. The procedure starts with reverse transcription/copying (*see* **Note 12**) of the RNA/DNA into cDNA, using tagged random octamer primers. It continues with the second strand synthesis by the same primer. The reaction is then purified from primers and single-stranded nucleic acids (*see* **Note 13**). The process ends with single-primer amplification of the random fragments by using the tag sequence. The principle is illustrated in Fig. 1. The final products are visualized on a 2% agarose gel.

3.2.1 First Strand Synthesis

Work in strips with lids or plates with foil. Whenever possible, use multichannel pipettes to speed up the process. Work on ice!

(A) *Procedure for RNA and DNA Viruses*

1. Prepare RT1 mix for the total count of samples + excess. For each sample prepare 1.0 μL of RNasin, 1.0 μL of 10 mM dNTP, 1.0 μL of 20 mM primer D2_8N, and 6.0 μL of H_2O (DEPC).
2. Prepare RT2 mix for the total count of samples + excess. For each sample prepare 4.0 μL of 5× first strand buffer, 1.0 μL of 0.1 mM DTT, and 1.0 μL of reverse transcriptase (200 U/μL).
3. Prefill 9 μL of RT1 mix into each position.
4. Add 4 μL DNA/RNA of sample and mix by pipetting.
5. Cover the plate/strip with lids or foil.
6. Denature in a thermocycler (at the pre-amplification area) at 65 °C for 5 min, followed by a hold at 4 °C.
7. Put the tubes on melting ice.

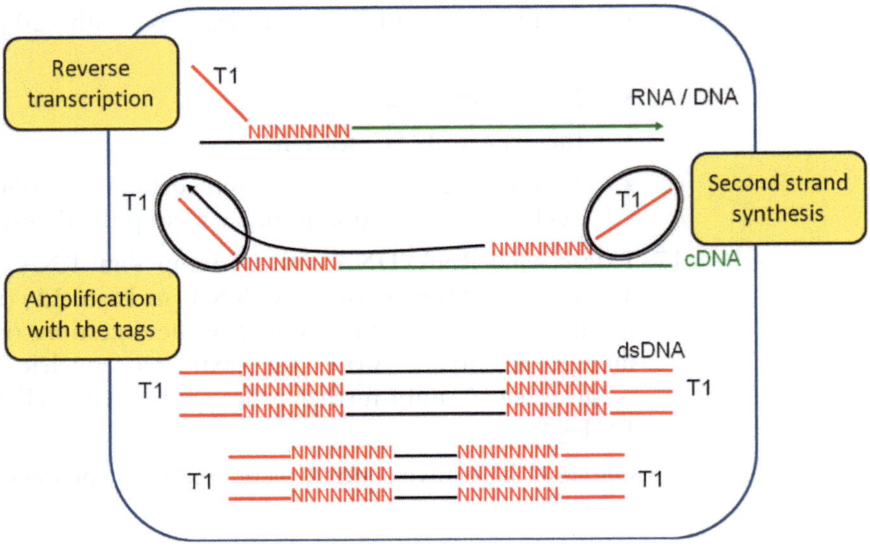

Fig. 1 The principle of random fragments formation and amplification

8. While on melting ice, add 6 μL RT2 mix into each well, mix by pipetting and briefly centrifuge.
9. Perform reverse transcription (on a cycler located at the pre-amplification area) at 25 °C for 20 min; followed by an incubation at 50 °C for 60 min; and then 70 °C for 15 min; and finally at 4 °C holding.

(B) *Alternative steps for DNA Viruses Only*

If no RNA viruses should be sequenced, the reverse transcriptase step has to be replaced with first strand synthesis using Klenow fragment.

1. Prepare KF mix for the total count of samples + excess. For each sample prepare 2.0 μL of 5× first strand buffer, 1.0 μL of 10 mM dNTP, 1.0 μL of 20 mM primer D2_8N, and 10.5 μL of H_2O (DEPC).
2. Prepare diluted Klenow fragment polymerase for the total count of samples + excess. For each sample prepare 1.1 μL of H_2O (DEPC). 0.15 μL of 10× Klenow buffer and 0.25 μL of Klenow fragment (10 U/μL).
3. Prefill 14.5 μL of KF mix into each well of the strip/plate.
4. Add 4 μL of the sample.
5. Denature in a thermocycler (at the pre-amplification area) at 95 °C for 5 min, followed by a hold at 4 °C.
6. When the cycler reaches 4 °C, place the reaction immediately on melting ice.
7. Add 1.5 μL of diluted Klenow fragment polymerase (i.e., 2.5 U enzyme) and mix well.

8. Perform first strand synthesis (on a cycler located in the pre-amplification area) at 37 °C for 1 h, followed by 75 °C for 10 min, and a final hold at 4 °C.

3.2.2 Second Strand Synthesis

1. Prepare diluted Klenow fragment polymerase (as above).
2. Denature the reaction from the previous step—this separates the newly synthesized tagged copy from the template. Use a thermocycler in the pre-amplification area: at 95 °C for 5 min, and a final hold at 4 °C.
3. Keep on melting ice.
4. Add 1.5 μL diluted Klenow fragment polymerase (i.e., 2.5 U enzyme) and mix by pipetting.
5. Perform the second strand synthesis (on a cycler located in the pre-amplification area): at 37 °C for 1 h; followed by 75 °C (*see* **Note 14**) for 10 min, and finally at 4 °C holding.

(SAFE STOPPING POINT: Here the reaction can be frozen.)

3.2.3 EXO—SAP Treatment

This assumes that the volume from the previous step is ~20 μL.

1. Prepare an EXO-SAP mix for the total count of samples + excess. For each sample prepare 17 μL of H2O (DEPC), 1.0 μL of 10× SAP Buffer, 1.0 μL of SAP, and 1.0 μL of Exonuclease I.
2. Add 20 μL of the EXO-SAP mix to the 20 μL reactions from the previous step.
3. Mix by pipetting; briefly spin.
4. Incubate in a pre-amplification area thermocycler, at 37 °C for 1 h, followed by 15 min at 75 °C and finally hold at 4 °C.

(SAFE STOPPING POINT: Here the reaction can be frozen.)

3.2.4 Single Primer Amplification

1. Prepare an amplification mix for the total count of samples + excess. For each sample prepare 26 μL of PCR grade water, 5.0 μL of 10× PCR buffer (AmpliTaq Gold), 6.0 μL of 25 mM MgCl2, 1.5 μL of 10 mM dNTP, 3.0 μL of 20 μM primer D2, and 0.5 μL of AmpliTaqGold polymerase (5 U/μL).
2. Prefill 42 μL of amplification mix into each well of the strip/plate.
3. Add 8 μL of the product from the previous step and mix by pipetting.
4. Bring the reactions into the post-amplification area—from this step onward, the work should be carried out in the post-amplification area (*see* **Note 15**).

5. Perform the single-primer amplification (on a post-amplification area cycler) using parameters: initial denaturation at 95 °C for 5 min, then 5 cycles at 95 °C for 1 min, 55 °C for 1 min and 72 °C for 1.5 min; then 25 cycles at 95 °C for30 s, 55 °C for 30 s and 72 °C for 1.5 min (+2 s with each cycle); followed by 72 °C for 10 min and finally at 15 °C holding.
6. Alternative amplification program—for higher viral loads (*see* **Note 16**): Initial denaturation: 95 °C for 5 min, followed by 20 cycles of 95 °C for 1 min, 55 °C for 1 min and 72 °C for 1.5 min; and a final synthesis of 72 °C for 10 min and finally at 15 °C holding.
7. Check the result of the reaction on a 2% agarose gel. There should be a smear without any noticeable distinct bands. The smear should extend from 100–200 bases to at least 400–500 bases. Use 2 μL of the product and mix it with the loading dye outside the original PCR plate.

Figure 2a shows an acceptable result, whereas Fig. 2b shows examples of failed samples.

3.3 Purification and Dilution Before Library Construction

In this section, the amplified fragments from the previous step are processed before the library is constructed: they are purified, size is limited, and samples are diluted to the correct concentration. Reasons for these steps are two-fold: (a) the protocol uses 2 × 250 base pairs sequencing kit which tends to fail with very short fragments, (b) tagmentation in the library construction is a random process, so it is advisable to use longer PCR fragments. First, the samples are subjected to a combined purification and size selection using the Ampure chemistry (*see* **Note 17**). Then the concentration of the purified sample is measured, and samples are diluted to 0.2 ng/μL, as the library construction using the Illumina Nextera XT kit requires 1 ng of input DNA in a volume of 5 μL.

3.3.1 Purification and Selection of Fragments 300 Bp+

Before the first use, make suitably sized aliquots of Agencourt AMPure XP beads.

Always start with the Ampure beads fully resuspended!

These steps can be automated on the Biomek series robots by Beckman.

Important: always use freshly prepared ethanol to ensure reproducibility.

1. Equilibrate a vial of AMPure XP beads at room temperature for approx. 20 min.
2. Take a new strip/plate and prefill 30 μL of each product. Store the rest of the PCR reaction as a back-up in the freezer and discard it after successful sequencing.
3. Shake the AMPure XP vial until magnetic particles are fully resuspended.

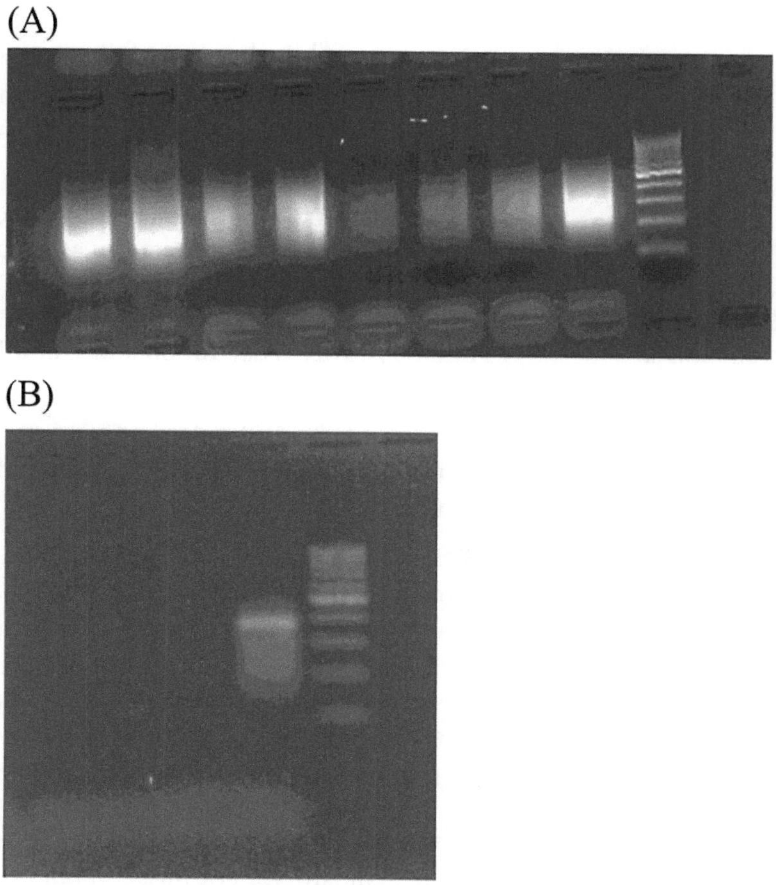

Fig. 2 Products of single-primer amplification. (a) Adequately processed samples (last position in each row is a molecular ladder (100 bp; its first stronger band is 500 bp). (b) Insufficient quality of processing: first two samples are too weak, the third has a discernible band. The molecular ladder is identical to Example 1

4. Add 15 μL Ampure to 30 μL of the sample. Pipette mix at least 10 times.
5. Incubate at room temperature for 5–10 min.
6. Place the reaction plate onto the magnet for 5–10 min to separate beads from the solution (until the sample is clear and all beads are on side).
7. Aspirate the supernatant from the reaction plate and discard.
8. Add 180 μL of 75% ethanol (freshly prepared) and incubate at room temperature for at least 30 s. Aspirate ethanol and discard.
9. Repeat for a total of two washes.
10. Aspirate the rest of ethanol with new dry tips.
11. Dry 10 min at room temperature (this is a maximum time, do not let the bead layer crack) (*see* **Note 18**).

12. Off the magnet: add 40 μL of PCR grade water, pipette mix 10 times (until the beads are fully resuspended in water).
13. Incubate at room temperature for 5 min.
14. Place the reaction plate onto the magnet for 10 min to separate beads from solution.
15. Transfer 30 μL of the purified product into a new strip/plate. Keep the same sample pattern as for previous plates.

3.3.2 Quantification of DNA Content

The quantification is performed using the Quant-iT™ dsDNA High-Sensitivity Assay Kit on and Qubit Fluorimeter.

1. Equilibrate the buffer and the reagent at room temperature (or use water bath for the buffer bottle). Reagent must be protected from light—then they are stable for at least 3 h.
2. Vortex all reagents and standards and briefly centrifuge.
3. Make a working solution by diluting Quant-iT dsDNA HS reagent (Component A) 1:200 in Quant-iT dsDNA HS buffer (Component B).
4. Prepare the assay tubes and mark them on lids (NOT on the side). One tube is used for each sample, plus two extra tubes are needed for calibration standards.
5. Prepare a dilution of the "0 ng/μL λ dsDNA HS standard" (Component C) to generate Standard #1, and a dilution of the "10 ng/μL λ dsDNA HS standard" (Component D) to generate Standard #2. For this, combine 190 μL of the above working solution and 10 μL of the respective standard.
6. In the tubes measuring samples, combine 195 μL of the working solution and 5 μL of samples.
7. Mix tubes by vortexing, briefly centrifuge, and let them incubate for 2 min, protected from light.
8. Follow the instructions on the Qubit screen, option "High sensitivity ds-DNA," starting with calibration of Standard #1 and Standard #2.
9. Measure the samples. The stock concentration (the one in the sample) could be counted by pressing "count stock concentration" with adjusting the calculation for 5 μL of the original sample volume in the reaction. NOTE: the result units are usually ng/mL (mililiter, not microliter).
10. Calculate and perform the dilution of the amplified samples to 0.2 ng/μL into PCR-grade water. Make the dilution in a new plate.

3.4 Library Construction Using the Nextera XT Kit

The Nextera XT kit fragments the source DNA and provides the ensuing fragments with adaptors and indices needed for downstream sequencing applications. The first step of the procedure is called "tagmentation." In this step an enzyme cuts the products and attaches a short sequence to the resulting fragments. The second step is a PCR-based adaptor and index attachment, which uses the short sequences that were attached to the fragments in the previous step. The principle is described and illustrated in detail in manuals to the kit.

3.4.1 Prepare Sample-Index Distribution

Illumina uses a dual indexing strategy—a sample is recognized by a unique combination of two indices. An N7 index is shared by all samples of a column, whereas an S5 index is shared by all samples of a row. The layout of indices for the index space of Nextera® XT Index Kit v2 Set A and of Nextera® XT Index Kit v2 Set D are shown in Annex 1 (the record sheet).

Special racks (*TruSeq Index Plate Fixture*) are available that hold the amplification plate, as well as both sets of indices (*see* Fig. 3). It is advisable to use the recommended strategy: **Columns 1–12** have index 1 primers ("N7..." primers)—here shown with red caps; **Rows A-H:** have index 2 primers ("S5..." primers) — here shown with white caps.

Keep a detailed record on the plate layout, and label each plate with the plate name and content. There are up to nine different steps in the procedure—each resulting in a new plate with a different kind of the product. All of them should keep the same sample layout (*see* Annex 1—the record sheet).

3.4.2 Tagmentation

This step fragments the DNA and provides the fragments with flanking sequences that later serve as targets for the indexing step. Work on ice unless advised otherwise.

1. Prepare from the Nextera XT DNA Sample Preparation Kit: ATM = Amplicon Tagment Mix, take it from the freezer and put on ice; TD = Tagment DNA Buffer, take it from the

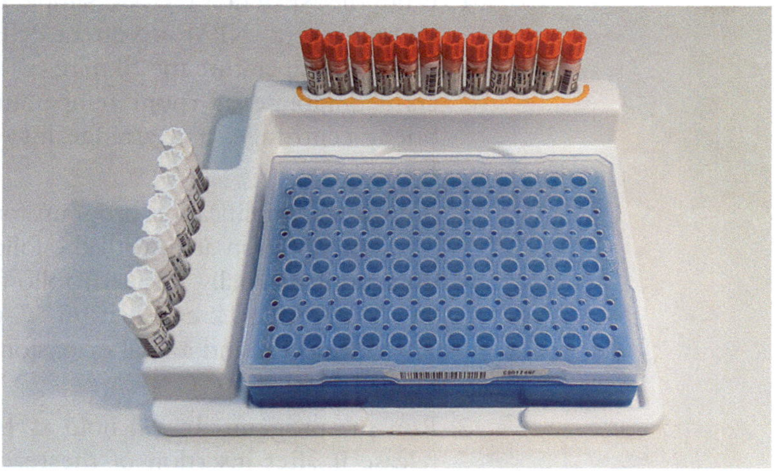

Fig. 3 TruSeq Index Plate Fixture with primer tubes

freezer and let it melt on the bench at room temperature; NT = Neutralizing Tagment Buffer, equilibrate to room temperature. If precipitate is present, resuspend it by vigorous vortexing. Buffers have to be mixed by vortexing, enzymes by inverting the tubes five times; then briefly centrifuged.

2. Amplified samples that have been diluted to 0.2 ng/μL: mix them by vortexing and briefly centrifuge.
3. Prepare a program for a thermocycler in the post-amplification area. Conditions are: 10 °C hold until manual continuation, followed by 55 °C for 5 min, and finally a hold at 10 °C.
4. Start it—it should hold at 10 °C and manually continue only after the samples are inserted into the prechilled block.
5. Prefill 5 μL of TD buffer to each well of a new PCR plate or strips.
6. To each well with TD buffer, add 2.5 μL of ATM.
7. Add 2.5 μL of sample (0.2 ng/μL). Mix well by pipetting, briefly centrifuge.
8. Insert into the cycler held at 10 °C and continue the program to perform incubation at 55 °C.
9. After the incubation at 55 °C, the temperature drops to 10 °C. Now take the samples out and stop the tagmentation by adding 2.5 μL of NT buffer to each sample (*see* **Note 19**) and mix by pipetting. This step is performed at room temperature!
10. Incubate on the bench at room temperature for 5 min. Briefly centrifuge.

3.4.3 Limited PCR for Indexing

This step adds adaptors with indices using a limited count of PCR cycles. Work on ice unless advised otherwise.

1. From the Nextera XT DNA Sample Preparation Kit and Index kit prepare: (a) NPM = Nextera PCR Master Mix, equilibrate at room temperature for 20 min; (b) N7.. and S5.. index primers—equilibrate at room temperature for 20 min. Vortex, briefly centrifuge, and arrange into the TruSeq Index Plate Fixture.
2. Prepare an amplification program for the thermocycler in the post-amplification area, with the following conditions: 10 °C, hold until manually released; followed by 72 °C for 3 min, 95 °C for 30 s; 12 cycles at 95 °C for 10 s; 55 °C for 30 s; 72 °C for 30 s; and a final extension at 72 °C for 5 min; and finally a hold at 10 °C.
3. Run the program. It will hold at 10 °C until you manually release it after insertion of the reactions into the prechilled block.

4. Buffers have to be mixed by vortexing, enzymes by inverting the tubes five times, then briefly centrifuged.
5. To a new PCR plate or strip, pipette 7.5 μL of NPM mix. Keep the same sample distribution as on the previous plate, respecting the layout of indices for your samples (*see* **Note 20**).
6. Pipette 2.5 μL of S5.. index primer and 2.5 μL of N7.. index primer to each position according to the Record Sheet. If possible, use a multichannel pipette and the Index Plate Fixture—this minimizes the risk of misplacing an index. Re-cap the tubes using new screw caps.
7. Add the whole volume of the completed tagmentation reaction (12.4 μL) and mix by pipetting. Use a multichannel pipette.
8. Cover the plate with foil or strips of lids, insert into the pre-chilled cycler block and launch the program. **(STOPPING POINT: After completion of the PCR, the library can be frozen.)**

3.5 Purification, Normalization, and Pooling of the Libraries

Libraries are now purified and their concentration measured. All libraries are then normalized to 500 pM (*see* **Note 21**) in order to equalize the share of their signal within the sequencing run. The first step is purification of samples with Ampure beads to remove unused PCR components. Then a small aliquot is diluted 1:10,000 and measured using a real-time PCR assay (KAPA kit which utilizes the sequencing adaptors on both ends of each fragment as priming sites) (*see* **Note 22**). Libraries are also checked on a Bioanalyzer instrument for the sequence length distribution. Each library is then diluted to 500 pM and then all libraries are pooled.

3.5.1 Cleanup with Ampure

As compared to the Ampure purification in **step 1** in Subheading 3.3, this step utilizes a different sample-to-Ampure ratio (1:1.8) and a different ethanol concentration (80%).

1. Equilibrate a vial of Agencourt AMPure XP beads at room temperature for approx. 20 min.
2. Shake the vial to fully resuspend magnetic particles.
3. Add 45 μL of AMPure XP to the 25 μL amplification reaction from the previous step. Mix by pipetting at least 10 times.
4. Incubate at room temperature for 5–10 min.
5. Place the reaction plate onto the magnet for 5–10 min to separate beads from the solution (unless the sample is clear and all beads are on side).
6. Aspirate the supernatant from the reaction plate and discard.

7. Add 180 µL of 80% ethanol (freshly prepared) and incubate at room temperature for at least 30 s. Aspirate the ethanol and discard.
8. Repeat for a total of two washes.
9. Aspirate the rest of ethanol with new dry tips.
10. Dry 10 min at room temperature (this is the maximum time, try not to let the bead layer crack otherwise the elution efficacy decreases).
11. Off the magnet: add 55 µL of PCR grade water, mix by pipetting 10 times (until the beads are fully resuspended).
12. Incubate at room temperature for 5 min.
13. Place the reaction plate back onto the magnet for 10 min to separate beads from solution.
14. Transfer 40 µL of the purified product to a new strip/plate. Discard the plate with beads.

3.5.2 Dilution Prior to KAPA Library Measurement

1. Prefill two deep-well plates (1 mL, DNA low bind plastic) with 495 µL of PCR grade water. Mark the first as "100×" and the second as "10,000×."
2. Add 5 µL of each purified library (product of the previous Subheading 3.5, **step 1**) to the corresponding well of the plate marked "100×."
3. Cover the plate with aluminum foil and mix by inverting 10 times.
4. Briefly centrifuge.
5. Transfer 5 µL from the "100×" plate to the "10,000×" plate.
6. Cover the plate with aluminum foil and mix by inverting 10 times.
7. Briefly centrifuge.
8. This 10,000-fold dilution of libraries will be used as a template for the library quantification.

3.5.3 KAPA Measurement and Quality Control on a Bioanalyzer Chip

1. If a kit is newly open, prepare primer-mix solution by adding 1 mL of 10× Illumina Primer Premix to the 5 mL vial of 2× KAPA SYBR® FAST qPCR Master Mix and mix well. Indicate date of primer addition clearly on the qPCR Master Mix vial.
2. Make a protocol with your library distribution and standards. Use at least duplicates of standards as well as of samples.
3. Prepare PCR mix for duplicates of the libraries plus standards: combine 6 µL KAPA SYBR® FAST qPCR Master Mix containing Primer Premix and 2 µL PCR water per each sample, plus excess.

4. Pipette 8 μL of PCR mix to each well of the detection plate for standards and samples.
5. Add 2 μL of standard (undiluted) or library (from the diluted "10,000×" plate).
6. Run absolute quantification real-time PCR with SybrGreen as the reporter (*see* **Note 23**): Conditions are as follows: denaturation at 95 °C for 5 min; 35 cycles at 95 °C for 30 s; and 60 °C for 60 s (with data acquisition in the SYBR green channel).
7. Based on the quantification results, calculate the concentration of the original (undiluted) libraries (i.e., multiply by the 10,000 dilution). The formula below also adjusts the concentration to the average length of the fragments:

   Undiluted library concentration =
   10,000* *Quantification result** 452
   / *Average library fragment length*

   (452 bp is the length of standards in the KAPA kit and 10,000 is the dilution factor; the average library fragment length is estimated from a gel photo, or from Bioanalyzer images).
8. Perform quality control by using High Sensitivity DNA Analysis Kit on a Bioanalyzer instrument. Assess the average fragment length for each sample. *See* Fig. 4 for an example.

3.5.4 Normalization and Pooling

1. According to the quantifications results, dilute all your libraries with PCR-grade water to a final concentration of 500 pM. Use a new plate. Mix well and briefly centrifuge.
2. Transfer 5 μL of each library into a common pool. Change tips to avoid cross-contamination (*see* **Note 24**).
3. This pool contains all libraries, and their *total combined concentration* is 500 pM. Label it clearly as "LIBRARY POOL," with the date of creation and name of sequencing run.
4. Store it at −20 °C in several aliquots until further processing (**SAFE STOPPING POINT**).
5. You may wish to use one of the aliquots for the verification of the concentration prior to denaturation and sequencing—this is done by using the same KAPA protocol as shown above.

3.6 Denaturation and Sequencing of the Pooled Libraries

The library pool will be now denatured and prepared for sequencing. This protocol uses a modified denaturation procedure in order to enable sequencing of lower concentrations of libraries than recommended by Illumina.

The input material for this section is the pool of libraries from the previous step. The total concentration of the pool is 500

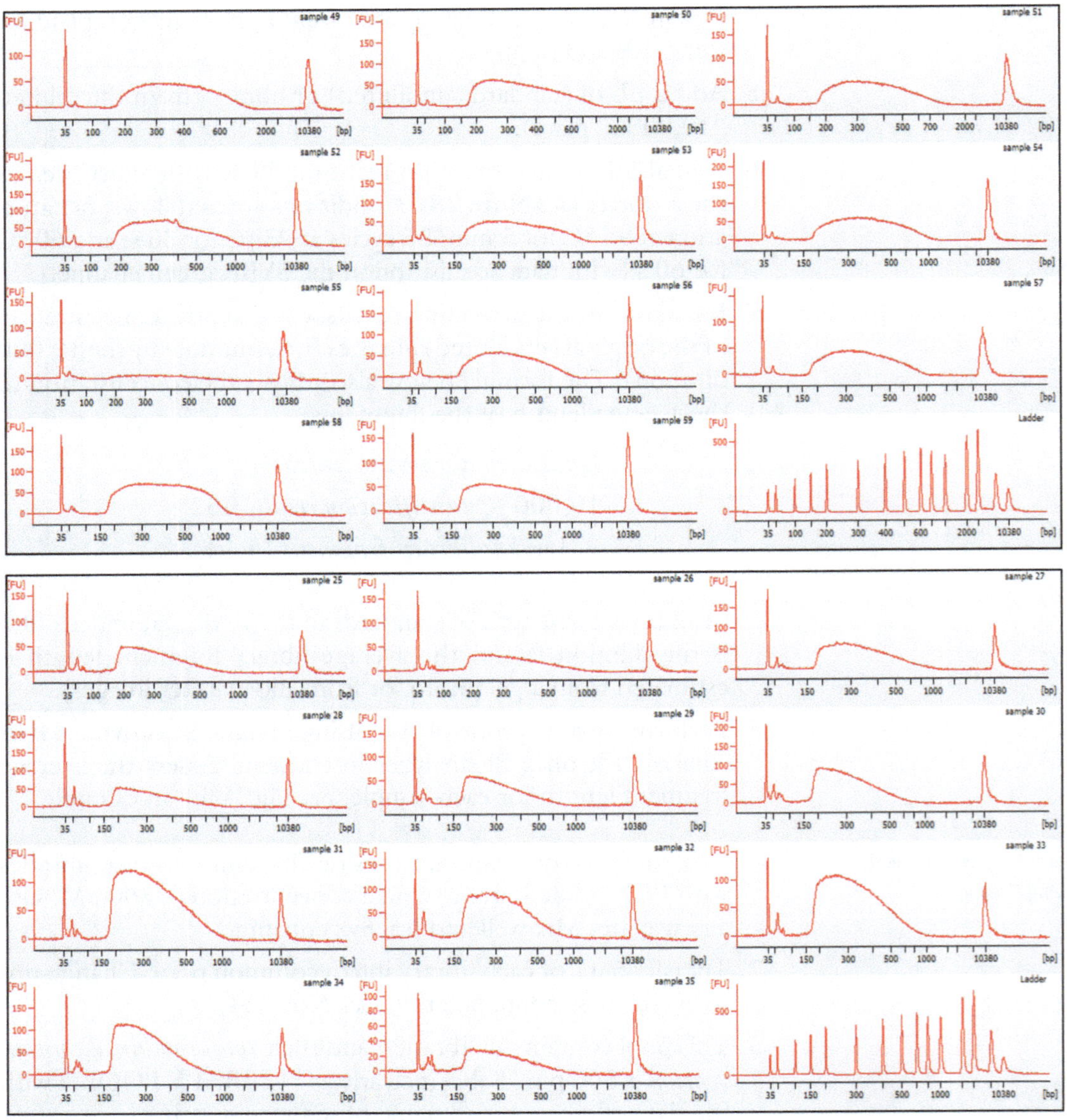

Fig. 4 Output from the 2100 Bioanalyzer—profiles of purified libraries. Both panels display 11 virome libraries, and one size marker (in the lower right corner). The library size distribution is smooth, with maximum between 300 and 500 bases, and has a clear left limit with a very small trace of unused primers (50 bp) next to the 35 bp marker

pM. The denaturation procedure consists of incubation with NaOH, followed by neutralization by HCl. The procedure is concluded with loading into the Illumina MiSeq sequencer. This denaturation procedure should be performed immediately before sequencing.

3.6.1 Denaturation

The starting total pool concentration is 500 pM. The final concentration of library used for sequencing will be 9 pM (*see* **Note 25**). The steps below should be performed *immediately before sequencing*.

1. Freshly dilute 0.2 M NaOH and 0.2 M HCl from concentrated stock solutions:
   - 0.2 M NaOH: first dilute the stock to 1 M (10 μL of 10 M NaOH plus 90 μL water). Then prepare a working 0.2 M solution (20 μL of 1 M NaOH plus 80 μL water).
   - 0.2 M HCl: first dilute the stock to 1 M (10 μL of 35% HCl plus 103 μL water). Then prepare a working 0.2 M solution (20 μL of 1 M HCl plus 80 μL water).
2. Preheat a heat block (1.5–2 mL) to 90 °C.
3. Melt HT1 buffer and place it on ice. Must be kept on ice at all times!
4. Vortex and briefly centrifuge the library pool from Subheading 3.5, **step 4**.
5. Prepare the MiSeq 2 × 250 bp kit:
   - Let the content of the sequencing cartridge melt in a water bath of *room temperature*. Water should not overreach the line on the cartridge. Let it melt for at least 1 h.
6. Fill in the Sample Sheet (*see* **Note 26**) using the Illumina Experiment Manager (IEM) (*see* **Note 27**):
   - Open by clicking on the icon.
   - Select: "Create Sample Sheet," "MiSeq," "Other," "FastQ Only".
   - In the Sample Sheet Wizard—Workflow parameters, select: "Nextera XT" (set A or set D according to index kit selected), "dual index reads," "paired end," "250" cycles read 1 and "250" cycles read 2. Then fill all other required fields with your run-specific information. Leave the Workflow-specific settings unchanged.
   - In the Sample Sheet Wizard—Sample selection: fill all required fields with your unique sample ID and index combinations.
7. Denature the sample pool:
   - Pipette 20 μL of the sample pool (500 pM) to a 1.5 mL tube marked as "POOL1."
   - Add 6.5 μL of 0.2 M NaOH.
   - Mix by pipetting and incubate in room temperature for 5 min.
   - Incubate in the heat block at 90 °C for 2 min.
   - Immediately place on melting ice and incubate for 5 min. Keep it on ice at all times.

- Add 600 μL of prechilled HT1, mix by pipetting and briefly centrifuge.
- Add 6.5 μL of 0.2 M HCl, mix by vortexing and briefly centrifuge. The concentration of the library is now 15.8 pM and the volume is 633 μL.

8. Dilute the denatured pool to 9 pM:
   - Transfer 340 μL of the POOL1 to a new 1.5 mL tube marked as "POOL2." Keep it on ice.
   - Add 260 μL of prechilled HT1.
   - Mix by vortexing and briefly centrifuge. The library is now diluted to 9 pM, denaturated and the NaOH has been neutralized. Keep it on ice, so that it does not re-anneal.
9. Denaturation of the PhiX control library:
   - To a 2 mL tube marked as "PhiX" add 2 μL of the PhiX control.
   - Add 3 μL of water.
   - Add 5 μL 0.2 M NaOH.
   - Mix everything by pipetting and incubate at 5 min in room temperature.
   - Dilute it with 1590 μL of prechilled HT1, mix by pipetting.
10. Mix denatured sample pool with denatured PhiX control library:
    - To a new 1.5 mL tube marked as "POOL+PhiX" combine 570 μL of the denatured and diluted library pool ("POOL2", 9 pM) and 30 μL denatured and diluted PhiX control (from the previous step).
    - Mix by vortexing, briefly centrifuge, and keep on ice.

3.6.2 Sequencing

Now sequence the mix of the libraries with PhiX ("**POOL + PhiX**") on an Illumina MiSeq instrument.

The bioinformatic analysis can then be accomplished, e.g., using VIPIE, a free online tool for parallel processing of multiple samples [11].

4 Notes

1. This protocol uses small 1 mL ultracentrifugation tubes. For other ultracentrifuges, the total volumes of samples will probably have to be adjusted according to the instrument manual, as most ultracentrifuges require having the tubes filled up. In such cases, adjust the sample volume with the HBSS buffer.

2. In principle, any random primer tagged with a unique sequence could be used after proper testing. This tagged primer was originally used in works by prof. E.L. Delwart and colleagues, Blood Systems Research Institute, San Francisco, CA [1, 2].
3. In particular, bacterial ribosomes tend to co-purify with viruses. The ribosomal fraction varies in size, it comes from bacteria disrupted naturally, or during thawing, centrifuging and filtering of the stool material. Bacterial ribosomal signal can then be subtracted upon bioinformatic analysis.
4. Optional: Inclusion of positive and negative controls is advisable, as it can control the enrichment and extraction procedure. A stool sample of known virus content, testable by quantitative PCR, can serve as a positive control. Alternatively, Armored RNA Enterovirus can be added into a negative sample, and then tested in the RNA extract using primers adopted from [12]. Negative control: HBSS.
5. If the filter gets stuck, use a new one for the left-over of the sample; then combine the flow-through of the two filters. If the unfiltered fraction is still high, use higher centrifugation speeds, but then the amount of contaminating ribosomal sequences increases.
6. Adjust the volumes according to the recommendations for a particular ultracentrifuge and tubes.
7. If ultracentrifuge is not available, this step may be accomplished by high-speed centrifugation on a conventional refrigerated centrifuge capable of min. 45,000 × *g*. The time of centrifugation must be then substantially prolonged in order to pellet the material (e.g., for 45,000–80,000 × *g* use at least 3 h at 5–8 °C).
8. The resulting pellet is seldom visible. Prior to centrifugation, make a label on the edge of the tube to localize the spot where the pellet should appear. Position the tube so that the mark on the edge faces outward the rotor.
9. You may wish to store the supernatant at −80 °C to enable repetition of the procedure in case the pellet is lost.
10. The use of enzymes can be detrimental to the total yield—pelleting by ultracentrifugation may have disrupted viral capsids. The effect of nucleases is never absolute—some ribosomal RNA always persists and it is often more effective to get rid of it by bioinformatic subtraction from the sequencing signal. A small amount of bacterial DNA is also introduced to the sample by PCR chemicals [13].
11. The Carrier RNA is a polynucleotide which will interfere with the downstream steps—ultimately, most bioinformatic tools will remove the uninterrupted runs of A's and T's, but the capacity of the sequencing will be compromised.

12. Reverse transcriptase not only transcribes RNA into cDNA, but also possesses a DNA-dependent DNA polymerase activity. Both activities are utilized in this protocol to simultaneously process RNA and DNA viruses. However, if only DNA viruses are to be sequenced, only Klenow fragment synthesis should be used instead of reverse transcription: this provides higher yields of DNA viral sequences (Subheading 3.2.1 Part (B)-Alternative steps for DNA viruses only). The protocol then continues with second strand Klenow fragment synthesis (Part 2.2).
13. The use of enzymes Exonuclease I (EXO) and Shrimp Alkaline Phosphatase (SAP) is efficient in cleanup of the reaction.

    EXO: Exonuclease I hydrolyzes single-stranded DNA in the $3' \rightarrow 5'$ direction, releasing 5′-mononucleotides and leaving the terminal 5′-dinucleotide intact.

    SAP: not only removes phosphate groups from the 5′-ends of DNA and RNA (which is useful in preventing re-ligation of linearized plasmid for cloning applications as well as end-labeling of probes), but also dephosphorylates dNTPs which allows for efficient removal of unincorporated nucleotides from PCR products.
14. Do NOT exceed 75 °C in order to avoid denaturation of the dsDNA product. Otherwise the product will be destroyed by exonuclease I in the subsequent step.
15. Separation of pre- and post-amplification area is extremely important in order to avoid a very serious, and potentially uncontrollable contamination by amplification products. Such contamination can arise not only from known agents, but also from yet unknown viral motifs—such a contamination would indeed be extremely difficult to detect and control.
16. The count of PCR cycles generally depends on the viral loads. Fewer cycles are recommended for high viral loads (especially when this protocol is used on viruses previously amplified on cell cultures). Generally, polymerases can introduce mutations, so the fewer cycles the less bias. Total number of 30 cycles generally works for any stool sample.
17. We strongly discourage using gel extraction: this almost inevitably leads to cross-contamination between samples that are processed in parallel!
18. Ethanol inhibits downstream reaction; its complete removal is thus warranted.
19. It is advisable to prefill the NT buffer to a strip of tubes, so that it can be quickly added using a multichannel pipette. This will eliminate uneven incubation times.

20. Indexing in next-generation sequencing is known to have certain limitations in its accuracy. Although dual indexing is rather accurate, in applications of virome sequencing it is still prone to contamination: during the sequencing process and demultiplexing, MiSeq machine sometimes assign the sequence to a wrong sample—if one sample contains a very high viral load, the virus then "emerges" in other samples. This signal cross-bleed has been published [10, 14]. Similarly, for genomic sequencing for monogenic disease diagnosis, it is recommended to confirm the results by specific PCR.

    This imperfection in indexing can be overcome by using a unique index combination for each sample, without sharing (reusing) the N or S indices within a single run. This is costlier, because it requires two set of indices (set A and set D) and allows parallel sequencing of only 16 samples in one run. The strategy of distribution of samples into the index space is illustrated here: unique indexing (Set A indexes in green, Set D indexes in orange, Sample with both indexes unique—gray).

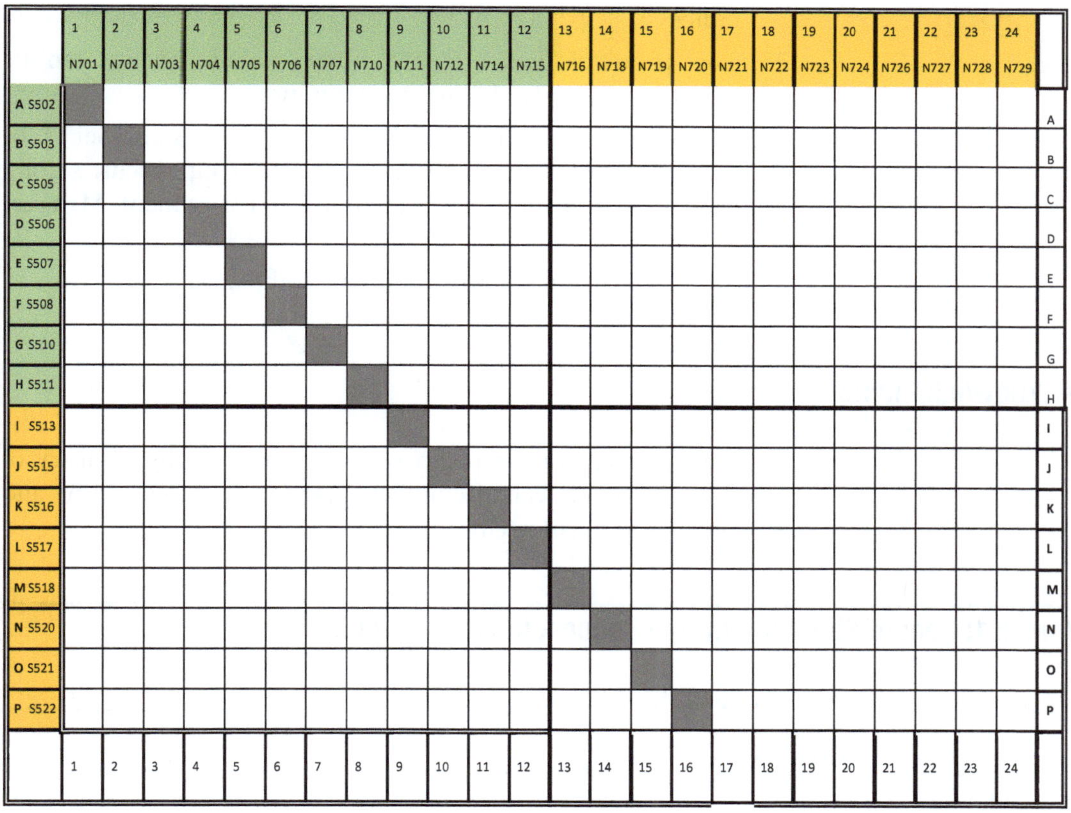

21. Virome library preparation usually results in quantities ranging within 500–4000 pM. Illumina MiSeq denaturation procedure requires at least a 2000 pM concentration—this is however not always possible in the virome library preparation, probably due

to variations in the input materials. Therefore, the denaturation protocol has been modified.

22. Although using this lengthy procedure with real-time PCR may seem redundant, in our hands this is far superior to any other library normalization strategies (e.g., commercially available plates with limited binding capacity, or bead-based normalization).
23. Consult the manual to the kit, as well as to your PCR cycler for instructions on analysis of very early amplification. Some real-time instruments obtain baseline levels of fluorescence during early cycles of PCR, e.g., 3–15, before usual applications generate amplification signals. However, the KAPA quantification of libraries generates extremely early threshold cycles, and the baseline collection intervals therefore must be set manually (set to, e.g., cycles 2–5). Otherwise the curves bend downward and the quantification is inaccurate.
24. Although these measures against contamination are no longer critical—the libraries now have their own combinations of indices.
25. This protocol uses a 9 pM library—reads quality can be adjusted with lower library concentration.
26. It is possible to combine indices from the Set A and Set D, or use indexes of your own specific design. Your specific setting has to be uploaded into the Illumina Experiment Manager folders.
27. Freely downloadable from Illumina webpages.

Acknowledgments

The development of the protocols was supported by Ministry of Health of the Czech Republic: grants nr. 15-31426A and 15-29078A to Charles University in Prague.

Annex 1: Record Sheet for Virome Sequencing (page 1 of 2)

Name of the sequencing plate: ______________________________ ____________.

Date of processing: __________________ Processed by: __________________.

Index space layout: set AFC-131-2001

A set	1 N701	2 N702	3 N703	4 N704	5 N705	6 N706	7 N707	8 N710	9 N711	10 N712	11 N714	12 N715
A S502												
B S503												
C S505												
D S506												
E S507												
F S508												
G S510												
H S511												

Index space layout: set DFC-131-2004.

D set	1 N716	2 N718	3 N719	4 N720	5 N721	6 N722	7 N723	8 N724	9 N726	10 N727	11 N728	12 N729
A S513												
B S515												
C S516												
D S517												
E S518												
F S520												
G S521												
H S522												

The actual experimental plate layout: note both indices, and sample ID.

ind.N7.. ind.S5...smp.id:	1	2	3	4	5	6	7	8	9	10	11	12
A												
B												
C												
D												
E												
F												
G												
H												

Record Sheet - Virome Sequencing (page 2 of 2).

Name of the sequencing plate: ______________________
______________________.

Step	Resulting reaction content	Done date, time	By whom	Notes
3.1	Nucleic acid extracted			
3.2.1	First strand synthesized			
3.2.2	Second strand synthesized, after Exo/sap treatment			
3.2.4	Single primer amplification			
3.3.1	Purified product of the amplification			
3.3.2	Products of the amplification, diluted to 0.2 ng/μL			
3.4.2	Products of tagmentation			
3.4.3	Products of limited PCR for indexing (libraries)			
3.5.1	Purified libraries			
3.5.2	Diluted libraries for KAPA measurement			
3.5.4	Libraries normalized to 500 pM			
3.5.4	Pool of the normalized libraries			
3.6	Sequencing: Instrument ID: __________ Kit serial #: ___________ BaseSpace ID:___________			

References

1. Blinkova O, Kapoor A, Victoria J et al (2009) Cardioviruses are genetically diverse and cause common enteric infections in south Asian children. J Virol 83(9):4631–4641
2. Blinkova O, Victoria J, Li Y et al (2010) Novel circular DNA viruses in stool samples of wild-living chimpanzees. J Gen Virol 91(Pt 1):74–86
3. Donaldson EF, Haskew AN, Gates JE et al (2010) Metagenomic analysis of the Viromes of three north American bat species: viral diversity among different bat species that share a common habitat. J Virol 84(24):13004–13018
4. Kapoor A, Victoria J, Simmonds P et al (2008) A highly prevalent and genetically diversified Picornaviridae genus in south Asian children. Proc Natl Acad Sci U S A 105(51):20482–20487
5. Kapoor A, Victoria J, Simmonds P et al (2008) A highly divergent picornavirus in a marine mammal. J Virol 82(1):311–320
6. Li L, Pesavento PA, Shan T et al (2011) Viruses in diarrhoeic dogs include novel kobuviruses and sapoviruses. J Gen Virol 92(Pt 11):2534–2541
7. Li L, Shan T, Wang C et al (2011) The fecal viral Flora of California Sea lions. J Virol 85(19):9909–9917
8. Victoria JG, Kapoor A, Dupuis K et al (2008) Rapid identification of known and new RNA viruses from animal tissues. PLoS Pathog 4(9):e1000163
9. Victoria G, Kapoor A, Li L et al (2009) Metagenomic analyses of viruses in stool samples from children with acute flaccid paralysis. J Virol 83(9):4642–4651
10. Kramna L, Kolářová K, Oikarinen S et al (2015) Gut virome sequencing in children with early islet autoimmunity. Diabetes Care 38(5):930–933
11. Lin J, Kramna L, Autio R et al (2017) Vipie: web pipeline for parallel characterization of viral populations from multiple NGS samples. BMC Genomics 18(1):378

12. Honkanen H et al (2013) Human enterovirus 71 strains in the background population and in hospital patients in Finland. J Clin Virol 56(4):348–353
13. Spangler R, Goddard NL, Thaler DS (2009) Optimizing Taq polymerase concentration for improved signal-to-noise in the broad range detection of low abundance bacteria. PLoS One 4(9):e7010
14. Kircher M, Sawyer S, Meyer M (2012) Double indexing overcomes inaccuracies in multiplex sequencing on the Illumina platform. Nucleic Acids Res 40(1):e3

Chapter 7

NetoVIR: Modular Approach to Customize Sample Preparation Procedures for Viral Metagenomics

Nádia Conceição-Neto, Kwe Claude Yinda, Marc Van Ranst, and Jelle Matthijnssens

Abstract

The democratization of next-generation sequencing (NGS) technologies has enabled scientists to explore the diversity of microbial life in various ecological niches in an unpreceded depth. The role of viruses as a key player in health and disease is becoming increasingly clear. To address the need for an up scalable, reproducible protocol to purify RNA and DNA viruses from a sample, we describe our optimized method. The *N*ovel *E*nrichment *T*echnique *O*f *Vir*omes (NetoVIR) allows researchers to attain a fast, reproducible, and high-throughput sample preparation protocol for NGS gut viromics studies. With appropriate prior homogenization steps, this protocol can be extended to any biological samples.

Key words Viromics, Sample preparation, Next-generation sequencing, Virome

1 Introduction

The human gut virome consists of three fractions: (1) bacteriophages infecting resident gut bacteria, (2) viruses infecting the human host, and (3) viruses in transit that neither infect bacteria nor the human host (e.g., plant viruses). In recent years, the decrease in cost of DNA sequencing by means of next-generation sequencing (NGS) technologies has led to several metagenomics studies in humans [1–4]. However, little is known about the effects of the different virus-like particle (VLP) purification steps in the viral populations from a sample. Moreover, the absence of a conserved viral genes together with the high diversity of viruses in terms of shape, genome composition, virion, and genome sizes makes it challenging to develop a protocol to purify VLPs.

As the role of viruses in human health and disease is further explored, there is a need for a high-throughput, reproducible protocol to purify RNA and DNA viruses from various biological samples. To better understand the effects of different purification steps on various VLPs, we quantitatively analysed these effects using a

Andrés Moya and Vicente Pérez Brocal (eds.), *The Human Virome: Methods and Protocols*, Methods in Molecular Biology, vol. 1838, https://doi.org/10.1007/978-1-4939-8682-8_7, © Springer Science+Business Media, LLC, part of Springer Nature 2018

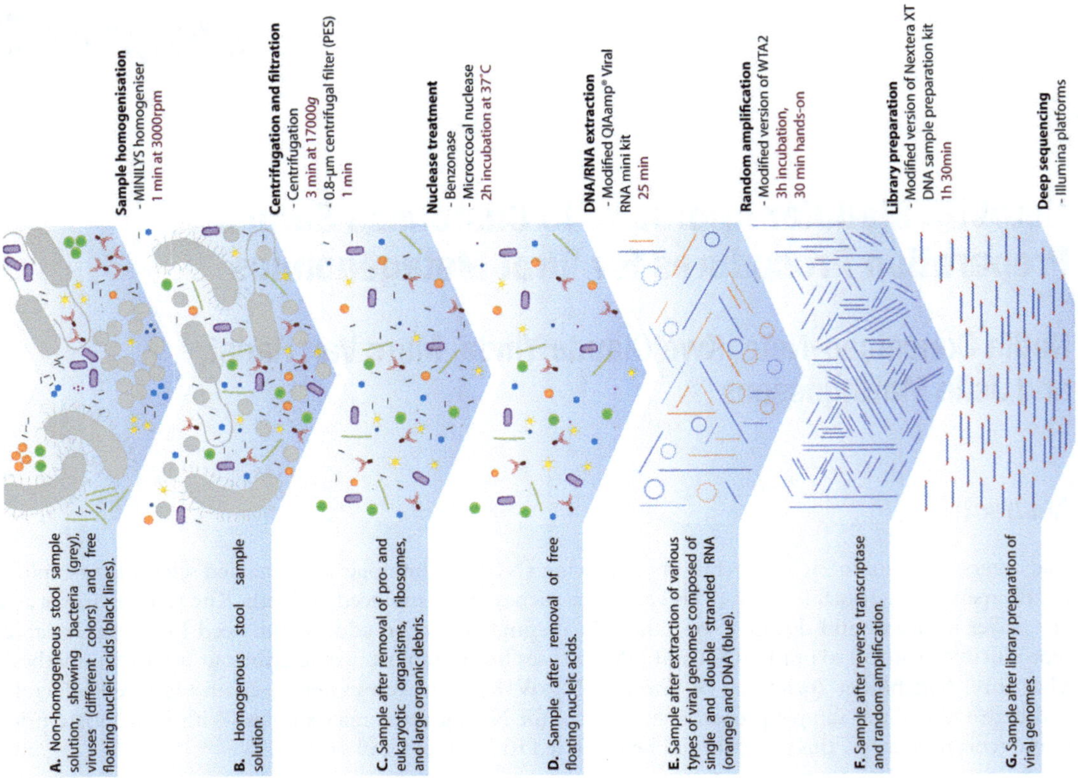

Fig. 1 Overview of the NetoVIR protocol. Each step (**a**–**g**) is explained and provided with an overview of the key materials/reagents and time needed for its conclusion

well-controlled viral mock community, allowing researchers to evaluate the bias introduced in sample purification [5]. We then selected a protocol that was able to strongly alter the ratio of viral versus bacterial and 16S rRNA genetic material in favor of viruses. Furthermore, our study indicated that most of the currently used virome protocols have overlooked giant viruses present in viromes, by making use of small filter pores and/or stringent centrifugation conditions [6]. Here, we describe the *N*ovel *E*nrichment *T*echnique *O*f *Vir*omes (NetoVIR, Fig. 1) that can provide a platform for reproducible and high-throughput sample preparation for NGS viromics studies.

2 Materials

2.1 General Stocks

1. Filtered tips (*see* **Note 1**).
2. Sterile 1.5/2 mL tubes.
3. PCR sterile tubes.
4. Vortex.

5. Microcentrifuge.
6. Gloves.

2.2 Sample Dilution

1. ~50–150 mg of faecal sample.
2. Laboratory scale.
3. Sterile Dulbecco's Phosphate-Buffered Saline.

2.3 Homogenization

1. Tissue homogenizer (*see* **Note 2**).
2. 2 mL empty tubes and caps with O-ring for tissue homogenizer.

2.4 Centrifugation

1. Benchtop centrifuge.

2.5 Filtration

1. Benchtop centrifuge.
2. 0.8 μm filter (PES membrane) (*see* **Note 3**).

2.6 Nuclease Treatment

1. Resolving enzyme buffer (100 mL): Add 12.11 g of 50 mM Tris, 1.47 g of 5 mM $CaCl_2$ and 0.61 g of 1.5 mM $MgCl_2$ to 80 mL of ultrapure water (prepared by purifying deionized water, to attain a sensitivity of 18 MΩ-cm at 25 °C) in a glass beaker. Then adjust the pH to 8.0 with HCl and add up to 100 mL. Filter the buffer through a 0.2 μm filter, aliquot and store at 4 °C.
2. Benzonase nuclease.
3. Microccocal nuclease.
4. Heating block (37 °C).
5. 0.2 M Ethylenediaminetetraacetic acid (EDTA).

2.7 Nucleic Acid Extraction

1. Total viral RNA and DNA extraction kit (*see* **Note 4**).
2. Benchtop centrifuge.
3. Microcentrifuge.

2.8 Random Amplification

1. WTA2—Complete Whole Transcriptome Amplification Kit.
2. Thermocycler.

2.9 PCR Product Purification

1. MSB Spin PCRapace PCR purification kit (*see* **Note 5**).

2.10 Concentration Measurement

1. dsDNA measuring assay device.
2. dsDNA high sensitivity measuring kit and tubes (*see* **Note 6**).

2.11 Library Preparation

1. Nextera XT DNA Library preparation kit (*see* **Note 7**).
2. Nextera Index kit.

3. Magnetic beads for purification.
   - Magnetic stand.
4. Thermocycler.
5. 80% ethanol freshly prepared using 100% ethanol and sterile H_2O.
6. Resolving 10 mM Tris–HCl, pH 8.0 buffer (1000 mL): Add 1.211 g of Trisma base (121.14 g/mol) to 800 mL of ultrapure water (prepared by purifying deionized water, to attain a sensitivity of 18 MΩ-cm at 25 °C) in a glass beaker. Then adjust the pH to 8.0 with HCl and add up to 1000 mL. Aliquot and store at room temperature.

2.12 Concentration Measurement

1. dsDNA measuring assay device.
2. dsDNA high sensitivity measuring kit (*see* **Note 6**).

2.13 Fragment Distribution Analysis

1. Automated electrophoresis (e.g., Bioanalyzer with the High Sensitivity DNA kit, *see* **Note 8**).

2.14 Quantitative PCR

1. 10 mM Tris–HCl, pH 8.0.
2. Real-time PCR instrument.
3. qPCR plates.
4. KAPA DNA Library quantification kit Illumina.

2.15 Sample Pooling

1. Sample libraries.
2. Ultrapure water.

3 Methods (*See* Note 9)

3.1 Enrichment of Faecal Samples for Virus-like Particles

1. Weigh the faecal material (≈50–150 mg) and add sterile PBS to attain a 10% (m/v) suspension.
2. Homogenize the sample using the tissue homogenizer at 3,000 rpm (speed 1) for 60 s (*see* **Note 10**).
3. Centrifuge the homogenized sample at 17,000 × *g* for 3 min and retrieve at least 200 μL of the supernatant for the next step.
4. Filter the supernatant obtained in a 0.8 μm filter at 17,000 × *g* for 1 min (*see* **Note 11**).
5. Add 7 μL of the premade 20× enzyme buffer to 130 μL of sample filtrate in an RNA-free 2 mL tube. To this mix add 2 μL of benzonase and 1 μL of micrococcal nuclease. Mix by gently inverting the tubes 3 times, spin briefly, and incubate for 2 h at 37 °C. Stop the reaction by adding 7 μL of 0.2 M EDTA and freeze samples at −80 °C or, preferentially, proceed for extraction.

3.2 Extraction

1. Add 560 μL of Lysis buffer to the RNase-free 2 mL tube with 140 μL of nuclease treated sample from **step 5**. Mix the mixture by pulse-vortex and incubate the mix at room temperature for 10 min. Change gloves.
2. Briefly centrifuge the mixture with a microcentrifuge and add 560 μL of absolute ethanol to the sample mixture. Mix very well by pulse-vortex and briefly centrifuge the mixture with a microcentrifuge. Change gloves.
3. Add 630 μL of the sample mixture to the spin column (*see* **Note 12**) and centrifuge at 6,000 × *g* for 1 min. Change the collection tube and repeat the procedure with remaining mixture.
4. Add 500 μL of wash buffer AW1 to the spin column. Centrifuge at 6,000 × *g* for 1 min and change the collection tube (*see* **Note 13**). Repeat the procedure with 500 μL of wash buffer AW2 and centrifuge at 20,000 × *g* for 3 min. Centrifuge again at 20,000 × *g* for 1 min and place the spin column in an RNase-free 1.5 mL tube.
5. Add 60 μL of AVE (elution buffer) to the spin column and incubate for 1 min (*see* **Note 13**). Centrifuge at 6,000 × *g* for 1 min and store viral RNA and DNA at −80 °C.

3.3 Amplification

1. To 2.82 μL of the extract add 0.5 μL of Library Synthesis Solution in RNase-free PCR tube.
2. Mix by pipetting and incubate in a thermocycler programmed for 95 °C for 2 min and then cool down to 18 °C.
3. To the cooled-primed RNA and DNA sample, immediately add the following mix (prepared on ice): 0.5 μL of library synthesis buffer, 0.78 μL of RNase-free water, and 0.4 μL of library synthesis enzyme.
4. Mix by pipetting and incubate in a thermocycler programmed at 18°C for 10 min, 25 °C for 10 min, 37 °C for 30 min, 42 °C for 10 min, 70 °C for 20 min, and 4 °C.
5. To the DNA library add the following mix (prepared on ice): 7.5 μL of amplification mix, 60.2 μL of Rnase-free water, 1.58 μL of WTA dNTP mix, 0.75 μL of amplification enzyme, and 5 μL of DNA library.
6. Mix by pipetting and incubate in a thermocycler programmed at 94 °C for 2 min, followed by 17 cycles of 94 °C for 30 s, and 70 °C for 5 min, and cooling down to 4°C (*see* **Note 14**).
7. Proceed to the next steps below with the amplified PCR product.

3.4 Purification of Amplified PCR Product

1. Add 500 μL of binding buffer to the 75 μL of WTA2 amplification product (*see* **Note 15**). Place the spin filter in a 2 mL receiver tube and add the 575 μL of mixture. Centrifuge at 11,000 × *g* for 3 min.
2. Place the spin filter in a 1.5 mL receiver tube, add 50 μL of elution buffer, and incubate for 1 min (*see* **Note 13**). Centrifuge at 11,000 × *g* for 1 min.
3. Store amplified DNA at −80 °C.

3.5 Quantification of Amplification Products

1. Set up the required number of 0.5 mL Qubit tubes for the two standards sand samples (*n* samples +2, *see* **Note 16**).
2. Prepare a working solution by adding per sample 199 μL of Qubit dsDNA HS Buffer and 1 μL of Qubit dsDNA HS Reagent (*see* **Note 17**).
3. Load 190 μL of working solution in the standard tubes and 199 μL in the tubes for samples.
4. Add 10 μL of standard to the respective tubes and 1 μL of sample in each sample tube.
5. Mix by pulse-vortex and incubate for 2 min at room temperature.
6. Read concentration values within 10 min of preparation for more accurate values.
7. Select the dsDNA High Sensitivity mode and run a new calibration. Read standard #1, standard #2 and the samples concentration. Calculate concentration according to the manual.

3.6 Library Preparation

1. Prepare a 1.2 ng/μL dilution of the WTA2 PCR product using Rnase-free water.
2. Make a mix of 5 μL of Tagment Buffer and 2.5 μL of Amplicon tagment mix in a 200 μL PCR tube. To this mix add 2.5 μL of 1.2 ng/μL dilution of the WTA2 PCR product from **step 6**, which was prepared on ice (*see* **Notes 18** and **19**).
3. Centrifuge the mix at 280 × *g* for 1 min (or using a microcentrifuge) and incubate in a thermocycler for 4 min at 55 °C, followed by cooling to 10 °C to ensure tagmentation. Proceed immediately to the next step.
4. Add 2.5 μL of Neutralize Tagment Buffer to stop the tagmentation process (*see* **Note 18**).
5. Add the following components to the tagmented DNA (12.5 μL): 2.5 μL of primer 1 (i7, N70X), 2.5 μL of primer 2 (i5, S50X), and 7.5 μL of Nextera PCR master mix.
6. Centrifuge the mix at 280 × *g* for 1 min (or using a microcentrifuge) and incubate in a thermocycler programmed at 72 °C for 3 min, followed by 15 cycles of 95 °C for 30 s, followed by

15 cycles of 95 °C for 10 s, 55 °C for 30 s, 72 °C for 45 s, and cooling down to 4°C (*see* **Note 20**).

7. Store amplified product from **step 4** DNA at 4 °C for a maximum of 24 h or proceed to PCR cleanup.

3.7 PCR Purification

1. Prepare fresh 80% ethanol.
2. Vortex the AMPure magnetic beads for at least 10 s.
3. Add 15 μL of magnetic beads to 25 μL of PCR product from **step 5** of Subheading 2.8.
4. Mix by pipetting 10 times to obtain a homogeneous solution.
5. Incubate this mix at room temperature for 5 min and place the plate in the magnetic stand for 2 min.
6. Carefully remove and discard the supernatant. Maintain the plate on the magnetic rack and add 200 μL of freshly prepared 80% ethanol. Incubate the plate/tube at room temperature for ≥30 s.
7. Repeat **step 9** once.
8. Remove all residual ethanol by pipetting and incubate the tube at room temperature for about 3 min without drying the beads (*see* **Note 21**).
9. Thoroughly resuspend the beads in 26.25 μL of Resuspension Buffer by pipetting them up and down 10 times. Incubate the plate at room temperature for 2 min and place the plate again on the magnetic stand for 2 min.
10. Carefully transfer 25 μL of the supernatant to new Rnase-free tubes and store the amplified DNA at −80 °C or proceed to the next steps (*see* **Note 22**).

3.8 DNA Library Quantification

Measure the concentration of each library by Qubit as described in **step 5** of Subheading 1 (*see* **Note 17**).

3.9 Fragment Distribution Analysis (See Note 23)

1. Remove all reagents from the fridge (4 °C) and allow them to reach room temperature for around 30 min before use. Keep dye and prepared gels from light.
2. Set up the chip priming station's lever to the lowest position and adjust the IKA vortex mixer to 2400 rpm.
3. Start the software, select assays: High Sensitivity DNA assay. Name your sample wells on the software.
4. Prepare the gel-dye mix by adding 15 μL of dye (blue-capped tube) to the red-capped tube. Vortex the dye mixture and transfer it to a spin filter, centrifuge for 10 min at 2240 × *g*, and discard the filter (*see* **Note 24**).

5. Pipette 9 μL of gel-dye mix in bold G well at the bottom to prevent air bubbles. To pressurise the gel-dye mix, place the chip in the priming station, lock it (click sound) and position the syringe on 1 mL, set the timer for 1 min. Press the plunger of the syringe down until it is held by the clip and after 1 min, release the clip, and make sure it moves up to at least 0.3 mL. Wait for 5 s or until it stops moving and pull the plunger to the 1 mL position and open the priming station. Add 9 μL of gel-dye mix in the remaining G wells.
6. Load 5 μL of marker (green-capped tube) in all sample wells and ladder well.
7. Load 1 μL of ladder (yellow-capped tube) in the ladder well.
8. Pipette 1 μL of sample in each 11 sample wells.
9. Vortex the chip for 1 min in the IKA vortex mixer.
10. Before and after every run, use a cleaning chip with 360 μL of Rnase-free water by placing it in the Bioanalyzer, close the lid, wait for a few seconds, and remove it.
11. Start your run within 5 min. Place the chip in the Bioanalyzer and close its lid carefully.
12. Check that the selection of dsDNA, High sensitivity DNA option is selected and the names of the samples are correctly filled in and start the run.

3.10 DNA Library Quantification by qPCR

1. Thaw all reagents from the kit. If the kit is used for the first time, add the 10× Primer Premix (1 mL) to the 2× KAPA SYBR FAST qPCR Master Mix (5 mL) and mix. Make a final qPCR Master Mix by adding 0.2 mL of the appropriate ROX (High or Low) depending on the qPCR instrument in your lab.
2. Prepare 1:1000 and 1:2000 dilutions of the libraries in 10 mM Tris–HCl, pH 8.0.
3. Prepare a master mix according to number of samples, six standards and appropriate replicates (*see* **Note 25**) by adding 12.4 μL final qPCR Master Mix and 3.6 μL Rnase-free water.
4. Pipette 16 μL of mix into each qPCR well and add 4 μL of standard or sample, accordingly.
5. Perform your qPCR with the following cycling conditions: 95 °C for 5 min, and 35 cycles of 95 °C for 30 s, and 35 cycles at 60 °C for 45 s.

3.11 Pooling of the Libraries

1. Dilute the library to 4 nM solution using the concentration obtained for the qPCR (**step 9** in Subheading 3) and the average size from the Bioanalyzer (*see* **Note 8**).

   If there are more than one sample, prepare a pool taking into account the number of reads desired per sample (*see* **Note 26**).

4 Notes

1. We have optimized this protocol with the MINILYS Tissue homogenizer. While another equivalent equipment can be used, we recommend comparing the homogenization speeds for optimal results, as more vigorous homogenization can lead to VLP rupture.
2. The use of filtered tips is required to avoid aerosol contamination between samples. It should be used along the full procedure or, if not possible, until the end of extraction.
3. The type membrane for filtration is crucial for an optimal procedure. We have optimized the protocol with the 0.8 μm filter PES viva clear mini (Sartorius).
4. This protocol was optimized using the QIAamp Viral RNA Mini kit (Qiagen). If you also make use of this kit proceed without addition of carrier RNA to the lysis buffer at the first step of extraction.
5. For the purification, a column PCR kit is preferable for more reproducible results. We describe the MSB Spin PCRapace procedure for its high-throughput nature.
6. For dsDNA quantification, a fluorescent nucleic acid stain is preferable for extra accurate results. We describe the Qubit quantification procedure.
7. The use of the Nextera XT DNA preparation kit is recommended due to its unique feature of DNA shearing using transposomes. The use of library preparation kits using a linker ligation approach might fail due to issues with the cluster calling due to the presence of the WTA adapters. Moreover, the preparation time is suitable for high throughput.
8. The Bioanalyzer instrument is recommended as they have an automated electrophoresis assay. Moreover, their High Sensitivity DNA kit profiling yields better results as often DNA libraries are of low input and requires only 1 μL of DNA library.
9. During the whole procedure, it is crucial to keep all working spaces sterile. Use and change gloves regularly.
10. Tubes should not have any type of beads, this results in significant loss of particular viruses. Without homogenization filters are more likely to clog.
11. In case of clogging of the filter membrane consider further diluting the sample or replace the spin membrane with a new one and transfer the remaining supernatant. Alternatively, a pre-filtration step with a larger pore size of 10 μm could be performed.

12. When putting the sample in the spin columns, make sure the pipet does not touch the column rim.
13. Pipet the elution buffer directly in the rim without touching it with the pipet tip.
14. The number of cycles chosen depends on the input material. Typically, we recommend 17 cycles, however for samples with expected low viral DNA and RNA content, an increase to 20 cycles should be considered. Similarly, for samples with a high level of input material, 12 cycles can be used.
15. Binding buffer is a concentrate, 99.7% isopropanol needs to be added accordingly.
16. Concentration is measured using the Qubit dsDNA HS assay kit, which provides two standards. Tubes should be purchased separately.
17. We recommend preparing 10% extra of working solution to assure enough volume.
18. The input material, tagmentation time, number of cycles, and extension time were optimized for faecal samples (with an expected average fragment size of 350 bp). Input material was increased and tagmentation time was reduced to 4 min, however consider testing other conditions for optimal results.
19. Perform reaction on ice.
20. The number of cycles chosen (15) and the extension time depends on the input material and the optimal fragment size. In this study, since tagmentation time was decreased, PCR extension was increased to 45 s. Consider testing other conditions for optimal results and refer to library preparation section of the study of Picelli et al. [7]. Also, perform all mix in this step on ice.
21. Watch the beads and if you observe cracks, proceed for next step immediately. The time can be highly variable depending on the temperature and humidity inside the lab.
22. If beads are aspirated together with the supernatant, thoroughly resuspend the solution and re-incubate. The presence of magnetic beads in the final library can lead to downstream issues.
23. Size distribution varies according to the amount of DNA used, tagmentation time, and extension time. Therefore, these can be adjusted to improve the size distribution. Nevertheless, a peak between 250 and 500 bp should be observed.
24. Gel-dye mix should be kept for maximum 6 weeks within preparation and it is sufficient for five chips. Protect from light.

25. For more reproducible results, we recommend duplicates or triplicates of sample and standards.
26. Do not mix in the same pool samples with the same double indexing.

Acknowledgment

We would like to thank Lila Close for insightful tips and tricks.

References

1. Pride DT, Salzman J, Haynes M et al (2012) Evidence of a robust resident bacteriophage population revealed through analysis of the human salivary virome. ISME J 6:915–926
2. Kapoor A, Kumar A, Simmonds P et al (2015) Virome analysis of transfusion recipients reveals a novel human virus that shares genomic features with hepaciviruses and pegiviruses. MBio 6(5):e01466-15
3. Lim ES, Zhou Y, Zhao G et al (2015) Early life dynamics of the human gut virome and bacterial microbiome in infants. Nat Med 21:1228–1234
4. Hannigan GD, Meisel JS, Tyldsley AS et al (2015) The human skin double-stranded DNA virome: topographical and temporal diversity, genetic enrichment, and dynamic associations with the host microbiome. MBio 6(5):e01578-15
5. Conceição-Neto N, Zeller M, Lefrère H et al (2015) Modular approach to customise sample preparation procedures for viral metagenomics: a reproducible protocol for virome analysis. Sci Rep 5:16532
6. Norman JM, Handley SA, Baldridge MT et al (2015) Disease-specific alterations in the enteric virome in inflammatory bowel disease. Cell 160:447–460
7. Picelli S, Faridani OR, Björklund AK et al (2014) Full-length RNAseq from single cells using Smart-seq2. Nat Protoc 9(1):171–181.

Chapter 8

Viral Genome Isolation from Human Faeces for Succession Assessment of the Human Gut Virome

David Mayo-Muñoz

Abstract

Despite the important role of the microbiota in the human gastrointestinal tract (GIT) and its impact on life-long health, the successional process through which this microbial community develops during infancy is still poorly understood. Specially, little is known about how the amount and type of viruses present in the GIT, i.e., the virome, varies throughout this period and about the role this collection of viruses may play in the assembly of the GIT microbiota.

The patterns of taxonomic change of the GIT viral community can be analyzed in a birth cohort of infants during the first year of life. The present chapter presents a detailed protocol for the isolation and extraction of viral nucleic acids from collected human faecal samples, whole genome amplification (WGA) using phi29 DNA polymerase and preparation for sequencing through high-throughput 454 pyrosequencing. The sequencing data can be posteriorly used for taxonomic classification in order to establish the composition of the virome present in each sample and to assess the process of viral dynamics through time.

Key words Virome, GIT microbiota, Succession, Virus isolation, Viral genome extraction, Whole genome amplification, High-throughput sequencing

1 Introduction

The gastrointestinal tract (GIT) is a complex ecosystem which harbors a great number of microorganisms such as bacteria, archaea, viruses, and eukaryotic communities coexisting and maintaining homeostasis. Among this community, viruses are thought to play critical roles in the GIT and its ecology as they contribute to human health, acting in concert with the immune system, controlling invading pathogens or helping control local inflammatory [1]. They have a function in gut homeostasis as they may influence food web regulating microbial communities [2]. Moreover, phages show a strong influence on the diversity, abundance, and structure of bacterial communities, as they contribute to bacterial host mortality and genetic diversity through horizontal genetic transfer [3]. However, they can also negatively affect health by eliminating

Andrés Moya and Vicente Pérez Brocal (eds.), *The Human Virome: Methods and Protocols*, Methods in Molecular Biology, vol. 1838, https://doi.org/10.1007/978-1-4939-8682-8_8, © Springer Science+Business Media, LLC, part of Springer Nature 2018

probiotic strains [4] or introducing new phenotypic traits to the bacterial hosts, such as antibiotic resistance [5]. In spite of its importance, few studies about viral diversity in the human gut have been carried out [6, 7]. As the lack of adequate cultured-based methods and conserved genetic elements for taxonomic classification make the analysis and characterization of viral community dynamics complicated [8], metagenomic approaches, such as high-throughput sequencing, are opening new ways of assessing the viral diversity in a wider range of environments such as human faeces [9].

Interestingly, the alteration of functional diversity within the microbiota community can affect health status [10]. Therefore, many metagenomic studies have focused their efforts in the assessment of the functional diversity present in the GIT microbiota of healthy individuals, to detect potential deviations in individuals affected by different diseases [11, 12]. Nevertheless, the assessment on infants has remained underexplored even though childhood is the critical period for gut microbiota assembly with strong repercussions in digestive and motility maturation, metabolic, immune and brain development in early life, defining life-long health [13, 14]. In addition, few longitudinal studies limited to a few infants have been carried out [7, 15, 16] even though they are essential for the understanding of the microbiome community succession in the human gut during infancy. Through comprehending the gut colonization patterns in healthy infants with time, we will be able to produce better definitions of the deviations in this process that result in microbiota imbalances and disease, allowing the development of strategies for the formation of health-promoting microbiotas.

According to Vallès et al. [17], the GIT microbiota development during the first year of life reveals an incomplete successional process with two distinct colonization phases, separated by the introduction of solid foods to the infants' diets. Initially, the newborn's GIT is not sterile [18, 19] and the mode of delivery (vaginally or by caesarean section) affects the development of the gut microbiota in early life, being the newborn's microbial community closely related to the one encountered during birth [20]. When infants are fed only milk, the richness, diversity, and complexity of interactions among taxa tend to increase, finding mostly bacteria with genes that facilitate lactate utilization [21]. However, a strong decrease in richness is produced between the third and seventh months, after the introduction of solid foods, which promotes the growth of bacteria encoding genes to allow the utilization of a larger variety of carbohydrates, vitamin synthesis, and xenobiotic degradation. This phase is followed by a trend toward richness recovery afterward. These richness changes are mainly due to the loss and gain of rare genera [22]. On the other side, viral richness and diversity community decrease during the first years of life, being inversely correlated with bacterial trends in an age-dependant manner [16]. These contrasting results might indicate that bacterial communities turn more resistant to viral infection as they become more rich and complex and that the increase in bacterial abundance and shift in composition drives a shift in the virus community composition as well. Besides, a trend of convergence toward the taxonomic composition of the maternal bacterial microbiota and virome by the end of the year is observed, suggesting that the convergence in bacterial community composition may drive a concomitant convergence in the infant gut viral communities. Although this succession is incomplete at the 1-year and an adult-like gut microbiota is not established before 3 years of age [23].

The protocol in this chapter shows details for isolation of viral particles from human faecal samples, viral nucleic acid extraction, whole genome amplification using Phi29 DNA polymerase, visualization and quantification and preparation for high-throughput shotgun sequencing. The sample preparation and virus genome extraction has been adapted from Pérez-Brocal et al. [8]. The protocol was followed in a non-published study that explored the patterns of taxonomic change along time during GIT virome development, to complement the study of Vallès et al. [17], where the corresponding analysis with the bacterial gut community was developed. With this aim, a set of faecal samples from healthy infants throughout the first year of life was collected and virus sequences were obtained to characterize the phylogenetic composition and genetic repertoire of the virome present in each sample. Faecal samples were also collected from the mother in order to assess the progression of the infant's microbiota toward an adult-like state and to see the changes that occur during pregnancy. The diversity and structure of these viral communities in relation to those of the bacterial communities was also analyzed.

2 Materials

Prepare all solutions using ultrapure water (prepared by purifying deionized water, to reach a resistivity of 18 MΩ-cm at 25 °C) and analytical grade reagents. Prepare and store all reagents at room temperature, unless otherwise stated. Carefully follow all waste disposal regulations when disposing waste materials (*see* **Note 1**).

2.1 Viral Genome Extraction from Faecal Samples

1. 70% ethanol.
2. 50% RNA later/phosphate saline buffer solution (PBS).
3. Ice.
4. Centrifuge with temperature control.

5. 1000 μL, 200 μL, 20 μL, and 2 μL pipettes and tips.
6. 1.5 mL tubes.
7. 2 mL tubes.
8. 0.2 μm pore filters.
9. 0.45 μm pore filters.
10. 10 mL syringes.
11. DNase cocktail: 14 U Turbo DNase (2 U/μL), 25 U/mg Benzonase (25 U/μL). Store at −20 °C.
12. 2 U of RNase A (10 mg/mL, 50 U/mg). Store at 4 °C.
13. 1× DNase buffer. Store at −20 °C.
14. Thermomixer.
15. 0.5 M EDTA (Ethylenediaminetetraacetic acid), pH 8.
16. QiAmp viral RNA extraction kit (Qiagen): AVL buffer, carrier (store at −20 °C), 0.1 μm-pore filters and UV-radiate, columns, AW1, AW2 and AVE buffer.
17. Absolute ethanol, filter through 0.1 μm pore filters and UV-radiate.
18. RNase inhibitor. Store at −20 °C.
19. RQ1 DNase and RQ1 DNase buffer. Store at −20 °C.
20. Stop solution: EGTA (ethylene glycol-bis(β-aminoethyl ether)-N,N,N′,N′-tetraacetic acid).

2.2 Viral Genome Amplification

1. Illustra GenomiPhi V2 DNA Amplification Kit (GE Healthcare) (*see* **Note 2**): sample buffer, enzymatic mix and reaction buffer (*see* **Note 3**). Store at −20 °C.
2. 1000 μL, 200 μL, 20 μL, and 2 μL pipettes and tips.
3. 1.5 mL tubes.
4. Ultrapure water
5. Minispin.
6. Thermomixer.
7. TBE 1X buffer solution: 89 mM Tris, 89 mM boric acid, 2 mM EDTA, pH 7.6.
8. 0.8% agarose in TBE 1× buffer solution (*see* **Note 4**).
9. Molecular weight marker: GeneRuler, ready-to-use, 100–10,000 bp DNA ladder Mix; 0.1 μg/μL. Store at 4 °C.
10. Loading MIX: 50 μL loading buffer 3×, 0.2 μL Gel Red. Store at 4 °C.
11. Parafilm.
12. Ultraviolet (UV) transilluminator with photo documentation camera.

2.3 DNA Quantification by Picogreen®

1. TE 1×: 10 mM Tris–HCl, 1 mM EDTA, pH 8.0.
2. Picogreen® solution in DMSO (Dimethyl sulfoxide). Store at 4 °C in darkness.
3. Fluorimeter.
4. Cuvette for the fluorimeter.
5. 200 μL and 2 μL pipettes and tips.
6. 1.5 mL tubes.

2.4 Sample Preparation for Sequencing

1. TE 1×: 10 mM Tris–HCl, 1 mM EDTA, pH 8.0.
2. 200 μL, 20 μL, and 2 μL pipettes and tips.
3. 1.5 mL tubes.

3 Methods

Before starting and after finishing, clean with 70% ethanol to prepare a sterile surface. Work always having your samples on ice in order to preserve the nucleic acids.

3.1 Viral Genome Extraction from Faecal Samples

1. Measure the dry weight of the total faecal samples and determine its value in the Bristol scale [24]. Transfer the faeces (between 0.5 and 1 g) to two 2 mL tubes and weight them. Add PBS at 4 °C. Vigorously homogenize in vortex and/or manual agitation. One of the tubes can be stored at −80 °C for bacterial analysis or as a back-up. The other tube is used in the following steps for virus isolation and viral genome extraction (*see* **Note 5**).
2. Centrifuge the samples at 4000 × *g* for 5–10 min at 4 °C (*see* **Note 6**).
3. Transfer the supernatant to a new 2 mL tube and repeat the centrifugation step (*see* **Note 7**).
4. Finally, do a stronger centrifugation at 13,500 × *g* for 5–10 min. Transfer the supernatant to a new 2 mL tube and discard the pellet.
5. Proceed to serial filtrations using sequentially 0.45 μm and 0.2 μm filters (*see* **Note 8**).
   (a) Securely assemble the syringe with a 0.45 μm filter using a twisting motion (*see* **Note 9**) and remove the plunger from the syringe.
   (b) Insert the sample solution into the top of the syringe leaving space for the plunger. Reinsert the plunger, hold the assembled syringe and filter vertically into a new 1.5 mL tube by gently pressing down on the syringe plunger to push the sample through the filter (*see* **Note 10**).

Subsequently, repeat the process for filtering the sample through a 0.2 μm filter in a new 1.5 mL tube.

6. Treat the samples with DNases cocktail (*see* **Note 11**) and 2 U of RNase A in 1× Turbo DNase buffer. 7 μL of Turbo DNase +1.0 μL of Benzonase +4 μL of RNase A. The volume of Turbo DNase buffer to be added is 10% of the final volume: 68 μL for 600 μL of sample being the final volume 680 μL (*see* **Note 12**).
   Treat for 90–120 min at 37 °C. At the end, place the samples on ice.
7. Inactivate DNases using 0.5 M EDTA: add 20.4 μL for 680 μL of sample to a final concentration of 15 mM. Incubate for 10 min at 75 °C (*see* **Note 13**).
8. Nucleic acid extraction with QiAmp viral RNA extraction kit (*see* **Note 14**):
   (a) Add 4 V of AVL buffer to your sample: 2801.6 μL for 700.4 μL of sample.
   (b) Add the carrier, 1% of the AVL buffer volume: 28 μL for 700.4 μL of sample.
   (c) Vortex for 15 s.
   (d) Incubate at room temperature for 10 min.
   (e) Add 4 V of absolute ethanol: 2801.6 μL for 700.4 μL of sample.
   (f) Vortex 15 s.
   (g) Transfer the volume to a column (not more than 700 μL at once) and centrifuge at 6000 × *g* for 1 min. Discard the flow-through and repeat the step until all the volume has been transferred. At the end, discard the collection tube and transfer to a new one.
   (h) Add 500 μL of AW1 buffer to the column and centrifuge at 6000 × *g* for 1 min. Discard the collection tube.
   (i) Add 500 μL of AW2 buffer to the column and centrifuge at 20,000 × *g* for 3 min. Empty the collection tube and return to the column.
   (j) Centrifuge the empty column 1 additional min in order to remove the remaining AW2 buffer. Discard the collection tube (*see* **Note 15**).
   (k) Transfer the column to a new 1.5 mL tube and proceed to elute adding 40 μL of AVE buffer (*see* **Note 16**). Incubate for 2 min and centrifuge at maximum speed for 1 min. Repeat for a second 40 μL elution in a new 1.5 mL tube (*see* **Note 17**).
   (l) Add 0.5 μL of RNase inhibitor (*see* **Note 18**) and store samples at −80 °C.

9. DNase treatment (optional step, *see* **Note 19**):
   (a) Transfer 9 μL from the first elution to a new 1.5 mL microcentrifuge tube.
   (b) Add 0.1 V (1 μL) RQ1 DNase buffer and 1 μL of RQ1 DNase and incubate for 30 min at 37 °C.
   (c) DNase with 1 μL of Stop solution (EGTA) and incubate for 10 min at 65 °C.
   (d) Store samples at −80 °C.

3.2 Viral Genome Amplification

Use Illustra GenomiPhi V2 DNA Amplification Kit (*see* **Note 20**).

1. Prepare as many aliquots of 9 μL of sample buffer as samples are going to be amplified, including controls (*see* **Note 21**).
2. Label the tubes and add on each one 1 μL of the original DNA from the first elution of the extracted DNA (**step 8** on Subheading 3.1). For the positive control, add 1 μL of an isolated viral sample that it is known to amplify beforehand. For the negative control, add 1 μL of ultrapure water.
3. Incubate at 95 °C for 3–5 min (*see* **Note 22**).
4. Cool down to 4 °C on ice.
5. Prepare a master mix in a 1.5 mL tube, adding 9 μL of reaction buffer and 1 μL of enzymatic mix per sample (*see* **Note 23**).
6. Add 10 μL of the master mix to each tube of DNA with sample buffer from **step 2** (*see* **Note 24**) and vortex briefly.
7. Incubate at 30 °C for 2–3 h (*see* **Note 25**).
8. Inactive the reaction at 65 °C for 10 min. Store samples at −20 °C.
9. Confirm amplified products by gel electrophoresis:
   (a) Prepare a 0.8% agarose gel.
   (b) While the gel solidifies, prepare the samples in a tube or parafilm. Mix 0.5 μL of amplified product with 1 μL of Loading MIX and 3.5 μL of ultrapure water.
   (c) Add TBE 1× buffer solution as running buffer in the gel tank and place the gel once solidified.
   (d) Load 0.5 μL of DNA ladder mix and the prepared samples.
   (e) Run the electrophoresis for 30 min and a 110 V electric field.
   (f) Visualize gel under UV light, taking the proper precautions (*see* **Note 26**).
   (g) Take pictures of the results using the photo documentation camera.

3.3 DNA Quantification by Picogreen®

1. Temper TE 1× and Picogreen® few minutes before use. In the case of Picogreen®, keep on dark.
2. Mix 1 μL of sample with 24 μL TE 1× in a 1.5 mL tube.
3. Prepare a 1:200 dilution of Picogreen® in TE 1× in a 1.5 mL tube (*see* **Note 27**). Mix well by vortexing.
4. Add 25 μL of the diluted Picogreen® to the 25 μL of diluted DNA from **step 2**, mix well by vortexing, and incubate in darkness for at least 3 min.
5. Transfer 50 μL to the measurement cuvette and proceed to measure in the fluorimeter. Make two measures per sample and calculate its average.
6. Finally, calculate the amount of DNA present in the original sample. The results are given in ng/μL if the calibration curve is used (DNA2) (*see* **Note 28**). In order to convert into ng/μL of the original sample, calculate back by multiplying by 50, because the sample is diluted 50 times in TE 1× and Picogreen solution (1:50 dilution). Then, multiply by the volume that remains in the tube (*see* **Note 29**) for knowing the total amount.

3.4 Sample Preparation for Sequencing

The randomly amplified viral genomes are used for high-throughput shotgun sequencing. 600 total ng of sample is the optimal amount required for shotgun sequencing library, but at least 300 ng is needed (*see* **Note 30**).

1. Samples are prepared for sequencing by adding a barcode and pooling them in groups of four samples per run and all samples in a group must have the same amount of DNA. Therefore, adjust the amount of all samples in a group to the one with less amount (*see* **Note 31**).
2. Samples are sent for sequencing in a 454 pyrosequencing Genome Sequencer FLX titanium plus, on an eighth of a PicoTiterPlate device.

Afterward, the obtained viral sequences allow, through bioinformatic analyses, a taxonomic characterization in order to establish the composition of the virome present in each sample. In addition, statistical analyses can be performed in order to assess the process of viral dynamics through time by analyzing the taxonomic richness and diversity, as well as the clustering based on viral composition and abundance.

4 Notes

1. It is crucial for avoiding cross contamination to physically separate in different laboratories an area for extraction, an area for store and prepare PCR reagents, an area for adding them to

DNA samples and a post-PCR area for electrophoresis, quantification and preparation for sequencing. Moreover, each area has to have its own pipettes and tips. Besides, for having Rnase-free material, it is required to wash up pipets, with SDS 1% followed by H_2O_2 3%. Thermomixer is washed with HCL 0,1 N. Use always tips and tubes non-previously used.

2. This DNA amplification process is non-PCR based and uses the Phi29 DNA polymerase, and random hexamer primers for unbiased WGA. It is a multiple displacement amplification (MDA) which is used for linear DNA molecules, such as genomic DNA [25]. The MDA method is based on isothermal amplification and does not require changing the reaction temperature. It provides many improvements in respect to other amplification techniques, such as PCR. The Phi29 DNA polymerase possesses high processivity and fidelity, i.e., a lower error frequency [26], and this allows highly uniform amplification across the genome. Hence, it is possible to carry out the robust amplification of entire genomes from little amounts of DNA sample. Therefore, this method has been used in WGA and holds the potential to allow large-scale studies, such as sequencing-based genetic studies, executed directly from primary sample sources, like genomes isolated from faeces [6] or tissue cells [8].
3. The sample buffer contains random hexamers that anneal to the DNA. The reaction buffer contains additional random hexamers, nucleotides, salts, and buffer. The enzymatic mix contains the Phi29 DNA polymerase.
4. Dilute agarose at 0.8% in TBE 1× buffer solution. Bring it to the boil in order to completely dissolve all the agarose.
5. In case the extraction is not going to be executed subsequently, store the samples at −80 °C. It is preferable to extract samples in pairs, which makes the balance when centrifuging easier. If the samples do not have the same volume, balance them with PBS. If the samples are frozen, wait to be thawed.
6. Previously, cold down the centrifuge to 4 ° C for 5 min. Centrifugation optimizes viral load as only bacteria and faeces sediment.
7. Repeat this step until no visible pellet is present in the tube and transfer the supernatant to a new 2 mL tube after each centrifugation. The number of centrifugations needed varies from sample to sample. The pellet can be used for bacterial analysis.
8. Since the average size of a bacterium is above 0.5 μm and the size of an average virus is below 0.1 μm [27], only virus and, probably, mitochondrial DNA, are obtained although the risk of losing big viruses exists.

9. In order to minimize contamination of the syringe filter, the filter can be held in the original packaging while attaching the syringe.
10. Avoid pressing excessively as this could cause the filter housing to burst. If the back pressure increases significantly, change for a new filter as it may have become clogged and complete the filtration. Filtrate all the volume, do a short spin to collect it all if necessary.
11. Additional purification step in order to degrade unprotected nucleic acids of bacterial or eukaryotic origin.
12. The volume of sample to be treated can be adjusted to 600 μL, storing the remaining filtered sample at −80 °C. In the case that a different volume is going to be treated, the volume of the Turbo DNase buffer must be adjusted accordingly to the 10% of the final volume.
13. If the extraction cannot be carried out subsequently, store the sample at −80 °C, once as much.
14. Despite the name of the kit, this treatment coextract both DNA and RNA viruses. For isolation of RNA viruses, continue with **step 9** on Subheading 3.1.
15. This step is crucial because AW2 buffer is an inhibitor and it may reduce the yield in downstream processes.
16. Make sure that the elution buffer is added in the filter.
17. The first elution fraction is more concentrated than the second one. The latter is stored as back-up.
18. It is recommendable to immediately aliquot all the sample in small volumes (~9 μL) in order to not froze-defrost the whole sample many times.
19. This treatment degrades DNA, leaving only RNA viruses present. It is an optional extra step if RNA viruses are going to be identified as well, which is not covered by the present protocol.
20. Extracted viral genomes have to be enriched by WGA methods because the number of viral particles is often so low that the extracted DNA does not contain the minimal DNA amount required for the preparation of a high-throughput sequencing library [28].
21. It is essential to prepare a positive and a negative control of amplification in order to monitor the satisfactory performance of the process by having amplified fragments in the positive control and to confirm that there are no contaminants by not having any amplicons in the negative control, as it contains no template.

22. This step denaturalizes the DNA strands and it is preferable to incubate just for 3 min.
23. 0.5 μL of enzymatic mix is also enough for amplifying and it can be completed with 9.5 μL of reaction buffer.
24. Imperative to change tips each time to not cross-contaminate samples.
25. 2 h may be enough time for amplification, but if the amount when quantifying is too low, incubate for a longer time.
26. A band of the same size is expected in all samples, corresponding to the viral genomic DNA. If smears are present, they correspond to degradation during the extraction. In the case that a sample does not show any band, discard it and repeat the amplification or the extraction step.
27. For a 200 μL solution, mix 1 μL of Picogreen® with 199 μL of TE 1×. Date the tube and store the remaining at 4 °C. Discard after 3 days.
28. If the results are given in μg/L, divide first by 1000 to convert into ng/μL.
29. In case of saturation, the measured values would be too high (around 1000 μg/L). Therefore, a 1:10 or even 1:100 dilution in TE 1× is necessary for an accurate measurement. Consequently, the dilution ratio would be 1:500 or 1:5000, respectively, which has to be taken into account when calculating back the amount of original sample.
30. If a sample does not reach the required amount, repeat the amplification. If the problem persists, repeat the extraction step.
31. If the volume of sample taken is less than 10 μL, complete up to 25 μL with TE 1× in order to avoid evaporation.

References

1. Górski A, Weber-Dabrowska B (2005) The potential role of endogenous bacteriophages in controlling invading pathogens. Cell Mol Life Sci 62:511–519
2. Weinbauer MG (2004) Ecology of prokaryotic viruses. FEMS Microbiol Rev 28:127–181
3. Jagadish MN, Huang D, Ward CW (1993) Site-directed mutagenesis of a potyvirus coat protein and its assembly in Escherichia coli. J Gen Virol 74(Pt 5):893–896
4. Ventura M, Sozzi T, Turroni F et al (2011) The impact of bacteriophages on probiotic bacteria and gut microbiota diversity. Genes Nutr 6:205–207
5. Davis B, Waldor M (2002) Mobile Genetic Elements and Bacterial Pathogenesis. In: Craig N, Craigie R, Gellert M, Lambowitz A (eds) Mobile DNA II. ASM Press, Washington, DC, pp. 1040–1059
6. Kim M-S, Park E-J, Roh SW et al (2011) Diversity and abundance of single-stranded DNA viruses in human feces. Appl Environ Microbiol 77:8062–8070
7. Reyes A, Haynes M, Hanson N et al (2010) Viruses in the faecal microbiota of monozygotic twins and their mothers. Nature 466:334–338

8. Pérez-Brocal V, García-López R, Vázquez-Castellanos JF et al (2013) Study of the viral and microbial communities associated with Crohn's disease: a metagenomic approach. Clin Transl Gastroenterol 4:e36
9. Breitbart M, Hewson I, Felts B et al (2003) Metagenomic analyses of an uncultured viral community from human feces metagenomic analyses of an uncultured viral community from human feces downloaded from http://jb.asm.org/on December 8, 2013 by National Institute of technology and evaluation. J Bacteriol 185:6220–6223
10. Calo-Mata P, Ageitos JM, Böhme K, Barros-Velázquez J (2016) Intestinal Microbiota: First Barrier Against Gut-Affecting Pathogens. In: Villa T, Vinas M (eds) New Weapons to Control Bacterial Growth. Springer, Cham, pp. 281–314
11. Sommer F, Anderson JM, Bharti R et al (2017) The resilience of the intestinal microbiota influences health and disease. Nat Rev Microbiol 15:630–638
12. Collado MC, D'Auria G, Mira A, Francino MP (2013) Chapter 15 - Human Microbiome and Diseases: A Metagenomic Approach. In: Watson RR, Preedy VR (eds) Bioactive Food as Dietary Interventions for Liver and Gastrointestinal Disease. Academic Press, San Diego, pp. 235–249
13. Vandenplas Y, Huysentruyt K (2017) The microbiome in healthy children BT. In: Green RJ (ed) Viral infections in children, vol Vol. 1. Springer International Publishing, Cham, pp 183–191
14. Penders J, Stobberingh EE, Brandt PA et al (2007) The role of the intestinal microbiota in the development of atopic disorders. Allergy 62:1223–1236
15. Vallès Y, Gosalbes MJ, de Vries LE et al (2012) Metagenomics and development of the gut microbiota in infants. Clin Microbiol Infect 18:21–26
16. Lim ES, Zhou Y, Zhao G et al (2015) Early life dynamics of the human gut virome and bacterial microbiome in infants. Nat Med 21:1228–1234
17. Vallès Y, Artacho A, Pascual-García A et al (2014) Microbial succession in the gut: directional trends of taxonomic and functional change in a birth cohort of Spanish infants. PLoS Genet 10:e1004406
18. Jiménez E, Marín ML, Martín R et al (2008) Is meconium from healthy newborns actually sterile? Res Microbiol 159:187–193
19. Gosalbes MJ, Llop S, Vallès Y et al (2013) Meconium microbiota types dominated by lactic acid or enteric bacteria are differentially associated with maternal eczema and respiratory problems in infants. Clin Exp Allergy 43:198–211
20. Dominguez-Bello MG, Costello EK, Contreras M et al (2010) Delivery mode shapes the acquisition and structure of the initial microbiota across multiple body habitats in newborns. Proc Natl Acad Sci 107:11971–11975
21. Tanaka M, Nakayama J (2017) Development of the gut microbiota in infancy and its impact on health in later life. Allergol Int 66:515–522
22. Vaishampayan PA, Kuehl JV, Froula JL et al (2010) Comparative metagenomics and population dynamics of the gut microbiota in mother and infant. Genome Biol Evol 2:53–66
23. Yatsunenko T, Rey FE, Manary MJ et al (2012) Human gut microbiome viewed across age and geography. Nature 486:222–227
24. Lewis SJ, Heaton KW (1997) Stool form scale as a useful guide to intestinal transit time. Scand J Gastroenterol 32:920–924
25. Silander K, Saarela J (2008) Whole genome amplification with Phi29 DNA polymerase to enable genetic or genomic analysis of samples of low DNA yield. Methods Mol Biol 439:1–18
26. Paez JG (2004) Genome coverage and sequence fidelity of 29 polymerase-based multiple strand displacement whole genome amplification. Nucleic Acids Res 32:e71–e71
27. Bettarel Y, Sime-Ngando T, Amblard C et al (2000) A comparison of methods for counting viruses in aquatic systems. Appl Environ Microbiol 66:2283–2289
28. Stepanauskas R, Fergusson EA, Brown J et al (2017) Improved genome recovery and integrated cell-size analyses of individual uncultured microbial cells and viral particles. Nat Commun 8:84

Part III

Respiratory Tract

Chapter 9

Introduction to Techniques and Methodologies for Characterizing the Human Respiratory Virome

Alicia B. Mitchell and Allan R. Glanville

Abstract

There have been great advances in the methodologies available for the detection of respiratory viruses. Accompanying this, our knowledge surrounding the impact of these viruses has also made a great leap forward. We have come a long way from the once commonly accepted belief that the lower respiratory tract was sterile and that the detection of any microbial species must represent a breach in host defence and likely be associated with symptomatic infection. With the advent of molecular detection techniques and improvements in sequencing-based methodologies to make these tools more accessible and cost effective, we now know that there is an abundant and diverse ecosystem within the lower-respiratory tract. This chapter will outline the clinical impact of the human respiratory virome, techniques for sampling the lower respiratory tract, the evolution of the diagnostic tools available, and the current limitations in our instruments and knowledge in this area. The human respiratory virome is an exciting new area of research that will continue to grow with the aid of the methodologies outlined in the following chapters and the advent of even more efficient tools in the future.

Key words Microbiota, Metagenome, Virome, Respiratory viruses, Virome, Polymerase chain reaction, Sequencing

1 Overview of the Human Respiratory Virome

The human respiratory virome is an emerging area of research, as there is currently a very limited understanding of the importance of resident viral species in health and disease. This area has taken its origins from insights gained from the human microbiome project, where we have begun to understand the potential importance of the pulmonary microbiome in different disease states such as asthma, chronic obstructive pulmonary disease (COPD), and cystic fibrosis (CF). The complex interplay between resident viral species and the human immune system has been largely underestimated, however Virgin and collaborators [1] suggested that the widely-held understanding that viruses are obligate pathogens within the respiratory tract needed revision. Culture-based methodologies,

Andrés Moya and Vicente Pérez Brocal (eds.), *The Human Virome: Methods and Protocols*, Methods in Molecular Biology, vol. 1838, https://doi.org/10.1007/978-1-4939-8682-8_9, © Springer Science+Business Media, LLC, part of Springer Nature 2018

which were the mainstay of viral detection for decades, led to a limited appreciation of the range of viruses present within the respiratory tract. Since the advent of molecular-based methods including PCR, and recently, accessible sequencing methodologies, a large body of evidence regarding the ever-increasing diversity of viruses present within the human body has begun to emerge. The viral metagenome, or virome, is a collection of the total DNA and RNA sequences derived from a community. Research into the virome has allowed a greater understanding of the range of species present within each body system and how these interact with the host and other members of the microbiome.

A small conglomerate of research has demonstrated the large range of viruses and bacteriophages present within the airways, commencing with a study by Willner and collaborators [2] in 2009, where the bacteriophage populations within the lungs of individuals with CF were compared with healthy controls. Willner et al. found a core set of phages within the lungs of healthy individuals and hypothesized they were representative of a transient, external environment. Conversely, there was an additional core set of phages found within the lungs of individuals with CF. These bacteriophages appeared to represent a shared internal environment due to their association with resident pathogenic bacterial species such as *Pseudomonas aeruginosa*. This same group also went on to characterize the viral species present within two sets of lungs explanted from patients with CF, and showed distinct spatial heterogeneity of viral species within the different lobes. Increased pathological changes in different areas of the lung corresponded to decreased viral diversity and lack of bacteriophages [3]. The demonstration of clear spatial heterogeneity within diseased lungs further complicates research into the respiratory virome, and highlights differences between health and disease states. Techniques that only sample one lobe of the lung may not capture the full diversity of the lung virome.

There is limited research describing the role of the virome in other chronic respiratory diseases such as asthma and COPD. The majority of the research in the area of the respiratory virome thus far has focused on the differences between individuals with an acute viral illness and healthy controls. In children who were acutely unwell with a febrile illness, a much greater viral abundance and diversity was detected in the nasopharynx compared with non-febrile controls [4]. However, this study also routinely detected human rhinovirus on the nasopharyngeal swabs of children both with and without a febrile illness, indicating the presence of asymptomatic carriage, which in the clinical domain is a reminder of the need to assess the operating characteristics of every test and the very real risk of a false positive result. Rhinovirus has been shown to achieve asymptomatic persistence in many study cohorts [5–8].

Furthermore, this study provides evidence for the early development of a resident nasopharyngeal virome. Asymptomatic children from 2 to 36 months of age were included in this analysis, all of whom demonstrated a diverse community of viral species in the absence of an acute viral illness [4]. A further study in pediatric patients indicated a similar increase in viral abundance and diversity in children with acute respiratory tract infections, with many sequencing reads corresponding to *Paramyxoviridae*, *Coronaviridae*, *Parvoviridae*, *Anelloviridae*, *Orthomyxoviridae*, *Picornaviridae*, and *Adenoviridae* families. In comparison, healthy children demonstrated a more limited diversity of sequence reads, corresponding mainly to *anelloviruses*, some community acquired respiratory viruses and bacteriophages [9]. A major limitation to improving our current knowledge is the lack of reference databases for viral species. Due to little research in this area, a large number of viral sequences detected using these new methodologies remain uncharacterized and unclassified, thus restricting the interpretation of much of the data being collected.

1.1 Sampling the Lower Respiratory Tract

Until quite recently, the lower respiratory tract was regarded as sterile. Hence any organisms detected were assumed to be pathogenic. Since the advent of more sensitive techniques including polymerase chain reaction (PCR) and sequencing, a diverse lower respiratory tract ecosystem has been discovered encompassing a range of bacteria, fungi, viruses, and other archaea which exist within the lungs of healthy individuals. Characterization of the lower respiratory tract microbiome, including the virome, has been limited by sampling techniques, and the risk of contamination from both the nasal and oropharyngeal microbiotas. There are no methodological studies published characterizing the impact of upper respiratory tract viruses on lower respiratory tract samples thus far. Therefore, this section will present the data for the bacterial microbiome studies investigating respiratory tract sampling methods.

Many challenges are present when attempting to characterize the microbiome of the respiratory tract. The dominant methodologies used to sample the lower respiratory tract include bronchoalveolar lavage (BAL) obtained via bronchoscopy, and spontaneously expectorated or induced sputum. Sputum presents a less invasive method of sampling the lower respiratory tract but has an increased risk of oropharyngeal contamination due to passing directly through the upper respiratory tract when expectorated. However, a range of studies have shown that when used as a sampling technique in diseases such as CF, bronchiectasis, and COPD, the biological signals detected have been significantly and meaningfully associated with multiple measures including severity of illness, airway inflammation, antibiotic use, and risk of subsequent exacerbations [10–15].

Bronchoscopy is a more invasive option, but it may offer a lower rate of upper respiratory tract contamination considering that samples do not pass directly through the upper respiratory tract. The bronchoscope is passed either through the mouth or nose, and down through the oropharynx to sample the lungs. Once inserted, lavage fluid is collected directly from distal airways. A recent study [16] using serial BAL analysis demonstrated that sampling the lungs via bronchoscopy was not significantly confounded by the oral microbiome, consistent with previous serial bronchoscopy studies [17]. The evidence from these studies support the understanding that minimal contamination from the upper respiratory tract is present in lower respiratory tract samples when utilizing bronchoscopic techniques. Therefore, the use of either bronchoscopy or induced sputum may allow collection of representative lung virome samples.

Furthermore, investigation of the respiratory virome is associated with greater challenges than characterizing the bacterial component of the lung microbiome. The respiratory microbiome has a significantly reduced biomass compared with the gut microbiome. The viral component is even smaller. The lack of a universal viral molecular marker, such as 16S rRNA for bacteria and the ITS region in fungi, as well as the large diversity of viral elements makes virome research difficult and highlights the need for specialized research efforts and methodologies.

2 History of Culture-Based Methodologies for the Detection of Viruses

Classically, viral isolation was achieved through in vitro cell culture. This became the a standard operating procedure in the majority of diagnostic laboratories by the 1970s due to the availability of highly purified reagents and commercially prepared cell-lines [18]. Through microscopic examination, viral proliferation through cytopathic cell changes could allow detection and identification of a range of viral species. Cell culture has long been regarded as the "gold standard" for viral identification by which the accuracy of many other methods, such as antibody-based or molecular approaches, could be compared. Rapid cell culture techniques (such as shell vial or cluster trays) lead to a significant decrease in the amount of time required to identify viral pathogens, from 3 to 7 days for traditional cell culture to less than 48 h for rapid culture techniques.

Advantages of using cell-based methodologies include the ability to detect viruses in new or unexpected infections, the ability to determine the viability of viruses that are detected and the opportunity for growth of virus stocks to allow further study and serotyping. Viruses must be alive and replicating in order to cause cytopathic changes to cells in vitro, which may also give information about virulence and vigor of the individual virus strain detected [19]. However, there are many limitations to culture-based methods, including the conditions needed to maintain virus viability during transport to the laboratory after the specimen has been collected, and the culture conditions required to grow many viruses. Since the advent of molecular and sequencing methodologies, many other viral species have been discovered which had been completely missed by culture due to unfavorable conditions in the laboratory, inhibiting growth [20].

3 The Role of Polymerase Chain Reaction (PCR)

The introduction of molecular-based viral detection techniques revolutionized the field of viral diagnostics. Where viral identification was once expensive, time-consuming and often, negative, the use of PCR technology has facilitated a much simpler, rapid, and accurate undertaking. Importantly, the use of PCR technology has introduced a new level of sensitivity to field of viral diagnostics, and allows isolation and identification of a single virion. In the case of certain chronic respiratory diseases, such as COPD, this may be important, as studies have shown that a single virion may be sufficient to induce an exacerbation [21]. Furthermore, these molecular methods are advantageous due to the reduced need for special growth conditions and the obligation to keep large stores of cultured cells and media. In early preparation steps, genetic material is extracted from cells which can be stored below freezing in small vials for extended periods of time, thus reducing the need for large storage space.

PCR is a method which causes the exponential amplification of a known polynucleotide DNA sequence, which allows for detection by other biochemical assays. PCR requires knowledge of the genetic sequence of the particular viruses which are being targeted, as the reaction utilizes sequence-specific primers to amplify conserved gene regions of each virus to allow identification. These two oligonucleotide primers flank the genetic region that is to be amplified, and through cycles of heat denaturation of the double-stranded (ds) DNA, annealing of the primers to their complementary sequences, and extension of these sequences through the use of a heat resistant *Taq* DNA polymerase, the region of DNA is amplified. This is achieved through temperature cycles which allow optimum conditions for each of these processes to occur. The optimal temperature for the primer annealing is specific for each set of primers and will need to be determined experimentally as part of the reaction optimization steps. Since the resultant strands of DNA are also complementary to the primers, successful cycles of the PCR allow an exponential increase in the amount of DNA product produced [22].

In preparation for performing a PCR, cells in biological samples must be lysed and the nucleic acids extracted either manually using acid guanidinium thiocyanate-phenol-chloroform extraction (AGPC) or by using commercially prepared kits. For respiratory samples, RNA viruses are the most common constituent. Therefore, an RNA extraction must be completed and the samples will then need to be subjected to a reverse transcription or cDNA synthesis step before a PCR reaction can be carried out [23]. This is done using a master mix including equimolar concentrations of each deoxyribonucleotide (dNTP), random hexamers, a RNase inhibitor, a reverse transcriptase enzyme, and the extracted viral RNA. This step also requires heat changes to allow annealing of nucleotides, activation of the reverse transcriptase, and the final denaturation of the enzyme. Once the cDNA has been prepared, this can act as the base product for PCR reaction.

For conventional PCR, after the reaction has proceeded to completion, gel electrophoresis using an agarose or polyacrylamide gel is required for visualization of the products. These utilize an electric charge to separate the DNA fragments by size, which can be compared to the expected product size based on the primers used, to determine presence of a viral species. These may be viewed using a DNA-staining dye under UV light. Additionally, these isolated bands of DNA fragments can be excised from the gel relatively easily for down-stream applications such as strain typing [24].

3.1 Quantitative PCR

Real-time, quantitative PCR (qPCR) has also been developed which allows information regarding virus identification and viral load to be determined directly from the PCR assay, without the need for gel electrophoresis to visualize results. qPCR is based on the measurement of emitted fluorescence during the PCR reaction, to determine the amount of PCR product generated during each reaction cycle. This is normally due to the introduction of a fluorescently labeled oligonucleotide, which emits a fluorescent signal at certain points during the reaction. As a result, a reaction curve is generated, mapping the exponential phase of the reaction, which is needed to calculate the initial copy number of DNA fragments at commencement of the PCR. The elimination of post-processing steps has reduced the chances for cross-contamination of products, as well as allowing a significant reduction in cost and time for the technologist [25].

3.2 Multiplex PCR

Multiplex PCRs allow multiple different viruses and subtypes to be assayed in a single reaction. These have been shown to be more sensitive than culture in many instances, with a greatly reduced turnaround time, hence are time efficient for clinical settings [26, 27]. Moreover, the use of multiplex reactions greatly reduces the time taken to complete diagnostic testing while reducing the technologist's workload and the quantity of laboratory expendables utilized. Multiplex PCR has also allowed a significant increase in the number of dual viral infections diagnosed, as multiple species can all be detected with a single test which is capable of distinguishing specific viruses. This has provided the impetus for many studies into the role of dual viral infections, how they affect patients and whether certain groups are more at risk of negative outcomes as a result [28].

3.3 Luminex xTAG Respiratory Viral Panel

The Luminex xTAG Respiratory Viral Panel (RVP) Assay will be described in detail in the following chapter. It was one of the first commercial multiplex PCR assays for respiratory viruses to be approved by the USA Food and Drug Administration (FDA). It allows detection of 12 different respiratory viruses within a single reaction, including *Influenza A, Influenza A subtype HI, Influenza A subtype H3, Influenza B, Respiratory Syncytial Virus subtype A, Respiratory Syncytial Virus subtype B, Parainfluenza 1, Parainfluenza 2, Parainfluenza 3 virus, Human Metapneumovirus, Rhinovirus,* and *Adenovirus.* As noted, this is one of the only panels to also allow subtyping of identified influenza A viruses. For influenza viruses, this panel has a sensitivity of 96.4% and 95.9%, with a specificity of 91.5% and 96% for influenza A and B respectively, when compared to cell culture methods. This assay also has over 99% specificity for all three subtypes of parainfluenza virus [28, 29]. A further study compared the xTAG RVP with other commercially available multiplex respiratory virus PCR assays, and found that the xTAG had the greatest sensitivity and specificity at 100% and 91% respectively compared with culture, thus rendering it the most sensitive multiplex panel available on the market [30]. There are some inherent limitations to the use of PCR for the detection of respiratory viruses. Since the majority of respiratory viruses are RNA viruses with high mutation rates due to the low fidelity of the RNA-dependent RNA polymerase, PCR assays need regular monitoring and quality assurance to ensure that the primers are still specific for the viruses being assayed. Also, there is limited diagnostic applicability in using PCR assays that are not FDA approved. The increased sensitivity and specificity afforded through using these molecular assays risks the cost of decreased clinical relevance. A positive test may detect a single virion that may or may not be biologically viable, so it is a debatable point of whether the result is clinically meaningful particularly as PCR only detects nucleic acids rather than determining information about infectivity.

Despite these clinical caveats, the use of molecular-based methods, including PCR, has revolutionized the field of diagnostic virology and allowed the propagation of the understanding that viruses are ubiquitous in the respiratory tract of humans. The use of PCR in the research laboratory has instigated studies investigating the incidence of respiratory viruses in asymptomatic populations. A range of published literature demonstrates that viruses are often found in the lower respiratory tract of healthy individuals [7, 8, 31–33]. However, as PCR is limited by the need for specific primers, other viruses that may be present within the respiratory tract may be completely missed. Slowly, we have begun to understand that we have a diverse and abundant virome within the lower respiratory tract based on the increased accessibility of sequencing methodologies [34]. However, as previously discussed, quantitative PCR assays are very important in quantifying the relative viral loads of the different members of the respiratory virome after they have been detected utilizing sequencing methods. Both techniques are complementary, as they reveal slightly different types of information relevant to clinical practice. It is not enough just to know what viruses are present, we also need to know their relative abundance, as this determines the potential need for interventional therapies.

4 Sequencing-Based Methodologies

In 1977, a substantial breakthrough was made by Sanger which has forever changed our ability to determine DNA sequences. This was the discovery of the dideoxy technique which utilizes chemical analogues of dNTPs which lack the 3′ hydroxyl group (ddNTPs). Without this, the dNTPs are unable to form a bond with the 5′ phosphate of the next dNTP, causing chain extension to halt at this point. Mixing in radiolabeled ddNTPs as a fraction of the concentration of the standard dNTPs, these are incorporated into the growing DNA strand randomly causing extension to halt. If four parallel reactions are run, each containing an individual ddNTP base, this will result in the generation of strands of every possible length which can then be run on a polyacrylamide gel alongside autoradiography to determine the sequence for a particular length of DNA [35, 36]. Multiple improvements were made to this method over the following years. The radio-labeling was replaced with fluorometric-based detection allowing the reaction to proceed in a single vessel. Detection was improved through the use of capillary-based electrophoresis [37–41]. These changes resulted in the first commercially available DNA sequencing machines [42]. At the time, the sequencing technologies were limited by the availability of high concentrations of pure DNA. However, this became substantially easier with the development of PCR.

4.1 Next-Generation Sequencing

The next generation of sequencing was led by the pyrosequencing approach, whereby pyrophosphate synthesis was measured using a luminescent method. Here, template DNA is attached to a solid phase. Each nucleotide is washed through the system, and as they affix to the DNA bound to the solid phase, pyrophosphate will be synthesized through a two-enzyme process, which can be measured to infer sequence [43, 44]. This was a great improvement on the original Sanger sequencing, as natural nucleotides could be used, thus reducing the need for substantial modifications. Additionally, this reaction could be observed in real time, reducing the need for lengthy electrophoresis steps [43, 45, 46]. This technology was later licenced to 454 Life Sciences, which made some modifications to the preparation of samples subject to pyrosequencing. Libraries of DNA molecules are attached to beads which are subject to a water-in-oil emulsion PCR to coat each bead in a clonal DNA population. These beads are then washed over a plate, where each bead should end up in a single well and the dNTPs can be washed over these wells. The synthesis of pyrophosphate is measured for each individual well. This allowed sequencing of around 400–500 base pairs per well, multiplied by approximately 1 million wells per plate [47]. This technique allowed the mass parallelization of sequencing, which greatly increased the quantity of DNA which could be run in each reaction, while substantially reducing the amount of time taken.

Following this, the other major next-generation sequencing approach was pioneered by Solexa, later acquired by Illumina. In this method, complementary oligonucleotides are bound to a flow-cell, and adaptor-bracketed DNA molecules are washed over them. Clusters of clonal populations for each original DNA strand are produced by a solid-phase PCR [48, 49]. Following this, sequencing is achieved by a sequencing-by-synthesis method, whereby the dNTPs stop the immediate binding of the next nucleotide due to the fluorophore occupying the 3′ hydroxyl position. This must be cleaved before the next nucleotide can be bound. Modified dNTPs and DNA polymerase are washed over the flow-cell bound clusters in cycles, whereby the fluorophores can be excited via lasers in between each cycle to determine the next nucleotide in the sequence. Following this, the fluorophore in the 3′ position is enzymatically degraded before the next cycle commences [50]. The introduction of these next-generation sequencing techniques has allowed sequencing to become accessible in laboratories worldwide due to increased automation, ease of use, and cost-effectiveness.

PCR-based methods have been compared with next-generation sequencing for the identification of viruses found within the respiratory tract. A study of 16 children with lower respiratory tract infections (LRTI) investigated the detection of a causative community acquired respiratory virus using both PCR and sequencing methodologies. In all cases except one, the sample viral species was detected by both PCR and sequencing. In that single exception, enterovirus was detected by sequencing alone [51]. Therefore, comparable yields to PCR may be achieved using sequencing methodologies, however, sequencing is often not as sensitive as PCR due to the depth of sequencing required to detect small viral loads. Conversely, sequencing does not require specific primer sequences which will allow other viruses and subtypes to be detected which may not be expected.

These two next-generation sequencing methodologies described, are the mainstay of virome research today. The limited number of manuscripts published in this field have all utilized either the Roche 454 pyrosequencing platform or the Illumina Solexa platform [2, 4, 9, 52–54]. Sequencing has allowed a plethora of information regarding viral species present within the human respiratory tract to be gathered. However, it has presented other challenges. Due to the large output of sequencing data that is generated with the next-generation sequencing approaches, new techniques and computing strategies have had to be developed to organize and analyze all of this information. The parallelization of sequencing means that gigabytes of data are able to be generated in a single run. Pipelines have been designed to organize the output of a large number of short-read sequences, and allow comparison against reference databases. Identification of viral sequences relies on alignment and classification against publicly available taxonomic databases assembled from previous sequencing data. Sequences sharing a large percentage of homology (usually above 97% similarity) are grouped into operational taxonomic units (OTUs), which give information about genus and species level abundance within a given sample [55]. However, these databases are currently limited by the lack of research in the area of the respiratory virome. A large number of viruses found in the respiratory tract remain uncharacterized and unclassified at the present time [3].

5 Conclusions

The techniques available for identifying and characterizing viruses found within the respiratory tract have evolved greatly over the last 50 years. From cell culture as the original "gold-standard" for diagnosis of viral infections, to molecular-based techniques which revolutionized the clinical picture of respiratory viral infections and allowed same-day results to guide treatment, to finally, the latest next-generation sequencing techniques which have support the concept that a complex and diverse viral eco-system exists within the lungs of each human being. The next section will discuss the current latest advancements in viral diagnostics with clear methods for the use of the Luminex xTAG Respiratory Viral Panel Multiplex PCR and Direct Multiplexed Whole Genome Sequencing to reveal full viral genomic information. These techniques have paved the way for our current understanding of the Human Respiratory Virome, and its importance in human health and disease. As this area continues to grow, use of these techniques will become essential to guide insights into such an important area which has largely been underappreciated to date.

References

1. Virgin HW, Wherry EJ, Ahmed R (2009) Redefining chronic viral infection. Cell 138(1):30–50
2. Willner D, Furlan M, Haynes M et al (2009) Metagenomic analysis of respiratory tract DNA viral communities in cystic fibrosis and non-cystic fibrosis individuals. PLoS One 4(10):e7370
3. Willner D, Haynes MR, Furlan M et al (2012) Case studies of the spatial heterogeneity of DNA viruses in the cystic fibrosis lung. Am J Respir Cell Mol Biol 46(2):127–131
4. Wylie KM, Mihindukulasuriya KA, Sodergren E et al (2012) Sequence analysis of the human virome in febrile and afebrile children. PLoS One 7(6):e27735
5. Kaiser L, Aubert JD, Pache JC et al (2006) Chronic rhinoviral infection in lung transplant recipients. Am J Respir Crit Care Med 174(12):1392–1399
6. Jartti T, Lehtinen P, Vuorinen T et al (2004) Persistence of rhinovirus and enterovirus RNA after acute respiratory illness in children. J Med Virol 72(4):695–699
7. Howard LM, Johnson M, Williams JV et al (2015) Respiratory viral detections during symptomatic and asymptomatic periods in young Andean children. Pediatr Infect Dis J 34(10) 1074–1080
8. Heinonen S, Jartti T, Garcia C et al (2015) Rhinovirus detection in symptomatic and asymptomatic children: value of host transcriptome analysis. Am J Respir Crit Care Med. https://doi.org/10.1164/rccm.201504-0749OC
9. Wang Y, Zhu N, Li Y et al (2016) Metagenomic analysis of viral genetic diversity in respiratory samples from children with severe acute respiratory infection in China. Clin Microbiol Infect 22(5):458.e1–458.e9
10. Rogers GB, van der Gast CJ, Cuthbertson L et al (2013) Clinical measures of disease in adult non-CF bronchiectasis correlate with airway microbiota composition. Thorax 68(8):731–737
11. Rogers GB, Zain NM, Bruce KD et al (2014) A novel microbiota stratification system predicts future exacerbations in bronchiectasis. Ann Am Thorac Soc 11(4):496–503
12. Molyneaux PL, Mallia P, Cox MJ et al (2013) Outgrowth of the bacterial airway microbiome after rhinovirus exacerbation of chronic obstructive pulmonary disease. Am J Respir Crit Care Med 188(10):1224–1231
13. Cox MJ, Allgaier M, Taylor B et al (2010) Airway microbiota and pathogen abundance in age-stratified cystic fibrosis patients. PLoS One 5(6):e11044
14. Zhao J, Murray S, Lipuma JJ (2014) Modeling the impact of antibiotic exposure on human microbiota. Sci Rep 4:4345
15. Zhao J, Schloss PD, Kalikin LM et al (2012) Decade-long bacterial community dynamics in cystic fibrosis airways. Proc Natl Acad Sci U S A 109(15):5809–5814
16. Bassis CM, Erb-Downward JR, Dickson RP et al (2015) Analysis of the upper respiratory tract microbiotas as the source of the lung and gastric microbiotas in healthy individuals. MBio 6(2):e00037
17. Segal LN, Alekseyenko AV, Clemente JC et al (2013) Enrichment of lung microbiome with supraglottic taxa is associated with increased pulmonary inflammation. Microbiome 1(1):19
18. Hsiung GD (1984) Diagnostic virology: from animals to automation. Yale J Biol Med 57(5):727–733
19. Ogilvie M (2001) Molecular techniques should not now replace cell culture in diagnostic virol-

ogy laboratories. Rev Med Virol 11(6):351–354
20. Leland DS, Ginocchio CC (2007) Role of cell culture for virus detection in the age of technology. Clin Microbiol Rev 20(1):49–78
21. Seemungal T, Harper-Owen R, Bhowmik A et al (2001) Respiratory viruses, symptoms, and inflammatory markers in acute exacerbations and stable chronic obstructive pulmonary disease. Am J Respir Crit Care Med 164(9):1618–1623
22. Innis MA, Gelfand DH, Sninsky JJ et al (2012) PCR protocols: a guide to methods and applications, vol 1. Academic press, San Diego
23. Vanspauwen MJ, Wolffs PF, Franssen FM et al (2014) Comparison of three different techniques for the isolation of viral RNA in sputum. J Clin Virol 61(2):265–269
24. Aaij C, Borst P (1972) The gel electrophoresis of DNA. Biochim Biophys Acta 269(2):192–200
25. Klein D (2002) Quantification using real-time PCR technology: applications and limitations. Trends Mol Med 8(6):257–260
26. Templeton KE, Scheltinga SA, Beersma MF et al (2004) Rapid and sensitive method using multiplex real-time PCR for diagnosis of infections by influenza a and influenza B viruses, respiratory syncytial virus, and parainfluenza viruses 1, 2, 3, and 4. J Clin Microbiol 42(4):1564–1569
27. Mahony J, Chong S, Merante F et al (2007) Development of a respiratory virus panel test for detection of twenty human respiratory viruses by use of multiplex PCR and a fluid microbead-based assay. J Clin Microbiol 45(9):2965–2970
28. Mahony JB (2008) Detection of respiratory viruses by molecular methods. Clin Microbiol Rev 21(4):716–747
29. Merante F, Yaghoubian S, Janeczko R (2007) Principles of the xTAG respiratory viral panel assay (RVP assay). J Clin Virol 40Suppl 1:S31–S35
30. Balada-Llasat JM, LaRue H, Kelly C et al (2011) Evaluation of commercial ResPlex II v2.0, MultiCode-PLx, and xTAG respiratory viral panels for the diagnosis of respiratory viral infections in adults. J Clin Virol 50(1):42–45
31. Srinivasan A, Flynn P, Gu Z et al (2013) Detection of respiratory viruses in asymptomatic children undergoing allogeneic hematopoietic cell transplantation. Pediatr Blood Cancer 60(1):149–151
32. Munywoki PK, Koech DC, Agoti CN et al (2015) Frequent asymptomatic respiratory syncytial virus infections during an epidemic in a rural Kenyan household cohort. J Infect Dis. https://doi.org/10.1093/infdis/jiv263
33. Jansen RR, Wieringa J, Koekkoek SM et al (2011) Frequent detection of respiratory viruses without symptoms: toward defining clinically relevant cutoff values. J Clin Microbiol 49(7):2631–2636
34. Popgeorgiev N, Temmam S, Raoult D et al (2013) Describing the silent human virome with an emphasis on giant viruses. Intervirology 56(6):395–412
35. Sanger F, Nicklen S, Coulson AR (1977) DNA sequencing with chain-terminating inhibitors. Proc Natl Acad Sci U S A 74(12):5463–5467
36. Chidgeavadze ZG, Beabealashvilli RS, Atrazhev AM et al (1984) 2′,3′-Dideoxy-3′ aminonucleoside 5′-triphosphates are the terminators of DNA synthesis catalyzed by DNA polymerases. Nucleic Acids Res 12(3):1671–1686
37. Smith LM, Fung S, Hunkapiller MW et al (1985) The synthesis of oligonucleotides containing an aliphatic amino group at the 5′ terminus: synthesis of fluorescent DNA primers for use in DNA sequence analysis. Nucleic Acids Res 13(7):2399–2412
38. Ansorge W, Sproat B, Stegemann J et al (1987) Automated DNA sequencing: ultrasensitive detection of fluorescent bands during electrophoresis. Nucleic Acids Res 15(11):4593–4602
39. Ansorge W, Sproat BS, Stegemann J et al (1986) A non-radioactive automated method for DNA sequence determination. J Biochem Biophys Methods 13(6):315–323
40. Prober JM, Trainor GL, Dam RJ et al (1987) A system for rapid DNA sequencing with fluorescent chain-terminating dideoxynucleotides. Science 238(4825):336–341
41. Swerdlow H, Gesteland R (1990) Capillary gel electrophoresis for rapid, high resolution DNA sequencing. Nucleic Acids Res 18(6):1415–1419
42. Hunkapiller T, Kaiser RJ, Koop BF et al (1991) Large-scale and automated DNA sequence determination. Science 254(5028):59–67
43. Nyren P, Lundin A (1985) Enzymatic method for continuous monitoring of inorganic pyrophosphate synthesis. Anal Biochem 151(2):504–509
44. Hyman ED (1988) A new method of sequencing DNA. Anal Biochem 174(2):423–436
45. Ronaghi M, Karamohamed S, Pettersson B et al (1996) Real-time DNA sequencing using detection of pyrophosphate release. Anal Biochem 242(1):84–89

46. Ronaghi M, Uhlen M, Nyren P (1998) A sequencing method based on real-time pyrophosphate. Science 281(5375):363–365
47. Margulies M, Egholm M, Altman WE et al (2005) Genome sequencing in microfabricated high-density picolitre reactors. Nature 437(7057):376–380
48. Bentley DR, Balasubramanian S, Swerdlow HP et al (2008) Accurate whole human genome sequencing using reversible terminator chemistry. Nature 456(7218):53–59
49. Fedurco M, Romieu A, Williams S et al (2006) BTA, a novel reagent for DNA attachment on glass and efficient generation of solid-phase amplified DNA colonies. Nucleic Acids Res 34(3):e22
50. Turcatti G, Romieu A, Fedurco M et al (2008) A new class of cleavable fluorescent nucleotides: synthesis and optimization as reversible terminators for DNA sequencing by synthesis. Nucleic Acids Res 36(4):e25
51. Yang J, Yang F, Ren L et al (2011) Unbiased parallel detection of viral pathogens in clinical samples by use of a metagenomic approach. J Clin Microbiol 49(10):3463–3469
52. Lysholm F, Wetterbom A, Lindau C et al (2012) Characterization of the viral microbiome in patients with severe lower respiratory tract infections, using metagenomic sequencing. PLoS One 7(2):e30875
53. Young JC, Chehoud C, Bittinger K et al (2015) Viral metagenomics reveal blooms of anelloviruses in the respiratory tract of lung transplant recipients. Am J Transplant 15(1):200–209
54. Zoll J, Rahamat-Langendoen J, Ahout I et al (2015) Direct multiplexed whole genome sequencing of respiratory tract samples reveals full viral genomic information. J Clin Virol 66:6–11
55. Dickson RP, Erb-Downward JR, Martinez FJ et al (2016) The microbiome and the respiratory tract. Annu Rev Physiol 78:481–504

Chapter 10

Targeted Sequencing of Respiratory Viruses in Clinical Specimens for Pathogen Identification and Genome-Wide Analysis

Yu Yang, Shannon D. Walls, Stephen M. Gross, Gary P. Schroth, Richard G. Jarman, and Jun Hang

Abstract

A large number of viruses can individually and concurrently cause various respiratory illnesses. Metagenomic sequencing using next-generation sequencing (NGS) technology is capable of identifying a variety of pathogens. Here, we describe a method using a large panel of oligo probes to enrich sequence targets of 34 respiratory DNA and RNA viruses that reduces non-viral reads in NGS data and achieves high performance of sequencing-based pathogen identification. The approach can be applied to total nucleic acids purified from respiratory swabs stored in viral transport medium. Illumina TruSeq RNA Access Library procedure is used in targeted sequencing of respiratory viruses. The samples are subjected to RNA fragmentation, random reverse transcription, random PCR amplification, and ligation with barcoded library adaptors. The libraries are pooled and subjected to two rounds of enrichments by using a large panel of oligos designed to capture whole genomes of 34 respiratory viruses. The enriched libraries are amplified and sequenced using Illumina MiSeq sequencing system and reagents. This method can achieve viral detection sensitivity comparable with molecular assay and obtain partial to complete genome sequences for each virus to allow accurate genotyping and variant analysis.

Key words Targeted sequencing, Next-generation sequencing, Genotyping, Respiratory disease, Respiratory virus, Viral enrichment, Pathogen discovery, Virome

1 Introduction

In targeted DNA or RNA sequencing, a panel of oligonucleotide probes are used to capture nucleotide contents containing interested sequences, the enriched targets are sequenced using next-generation sequencing (NGS) to achieve sensitive detection and sequence-based analysis [1]. Several methods with comparable performance had been developed and successfully applied to

Electronic supplementary material: The online version of this chapter (https://doi.org/10.1007/978-1-4939-8682-8_10) contains supplementary material, which is available to authorized users.

Andrés Moya and Vicente Pérez Brocal (eds.), *The Human Virome: Methods and Protocols*, Methods in Molecular Biology, vol. 1838, https://doi.org/10.1007/978-1-4939-8682-8_10, © Springer Science+Business Media, LLC, part of Springer Nature 2018

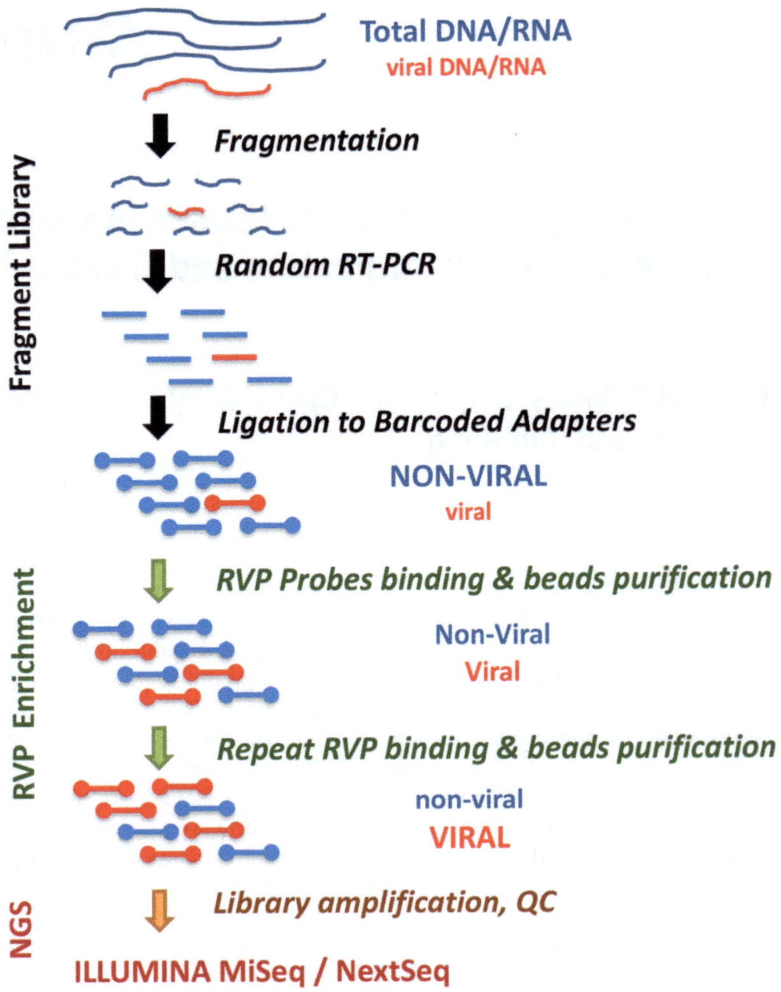

Fig. 1 Outline of the targeted sequencing of respiratory viruses in total nucleic acids from clinical specimens. The method uses TruSeq RNA Access Library Prep protocol and custom RVP oligos for genome-wide capture of 34 human respiratory viruses for next-generation sequencing (NGS)-based detections

sequencing of exomes, transcriptomes, cancer genes, significant pathogens, etc. [2–8]. Utilizing TruSeq RNA Access protocol and a large panel of oligos for whole-genome capture of 34 human respiratory viral pathogens (TruSeq RVP) enables enriching sequences of respiratory DNA and RNA viruses out of complex clinical specimens and production of genome sequences for genotyping and genetic variant analysis. The process includes collection, transportation, and storage of respiratory specimens, extraction of DNA and RNA, preparation of TruSeq fragment library, RVP enrichment, NGS and sequence data analysis (Fig. 1). The laboratory procedures require approximately 5 days from specimens to NGS run excluding specimen acquisition, NGS instrument run time and post-run data processing and analyses.

The method is expected to detect all commonly known respiratory viral pathogens that are diagnosed with molecular tests such as Luminex xTAG RVP assay, FilmArray Respiratory Panel (RP) tests, multiplex real-time PCR tests for respiratory viral infection with comparable sensitivity and accuracy [9–11]. The large capture panel and the scheme of genome-wide capture allow detection of most known respiratory viruses and viruses with high sequence divergences. The method is robust in handling specimens with viruses of a wide range of concentrations as well as specimens of compromised quality, e.g., degraded nucleic acids or high non-viral contents. The high capture efficiency and the superior sensitivity from this method make it a powerful tool for discovery of respiratory virome. However, there is an elevated risk of false positivity which demands more stringent contamination control procedure for the investigation of clinical specimens (*see* **Note 1**).

2 Materials

Perform the procedures in a Biosafety Level 2 (BSL-2) laboratory with a certified biosafety cabinet (BSC). Use clinical specimens and agents that can cause infection within the biosafety cabinet.

2.1 Nucleic Acids from Respiratory Specimens

Extract and purify DNA and RNA contents from respiratory swabs stored in a viral transport medium. Divide purified DNA/RNA into small aliquots, e.g., 10 μL per vial and store at −80 °C (*see* **Note 2**).

2.2 Library Preparation

1. Illumina TruSeq RNA Access Library Prep Kit (*see* **Note 3**). In this protocol, Coding Exome Oligos provided with the TruSeq RNA Access Library Prep kit are replaced with custom RVP oligos (*see* Supplementary file RVP capture probes for oligo sequences). The RVP oligos were designed to target whole genomes of 34 human respiratory viral pathogens (Table 1).
2. Illumina PhiX control V3 kit.
3. SuperScript III Reverse Transcriptase (200 U/μL) (*see* **Note 4**).
4. Agencourt AMPure XP beads. Store at 4 °C. Before use, equilibrate the beads at room temperature for 30 min, vortex vigorously for 1 min to thoroughly suspend the beads.
5. Nuclease-free water.
6. 80% ethanol (EtOH): Prepare fresh 80% EtOH and use within a day.
7. 0.2 M Tris–HCl, pH 7.0.
8. 10 nM Tris–HCl, pH 8.5 with 0.1% Tween 20.

Table 1
Respiratory viruses targeted by TruSeq RVP RNA access sequencing

Respiratory virus (34)	Genome size	Type	Target genome sequences
Respiratory syncytial virus B (S2)	15,190	(−)ssRNA	NC_001803.1
Respiratory syncytial virus A	15,225	(−)ssRNA	AY353550
Influenza A (H9N2)	13,500	(−)ssRNA	NC_004910.1 NC_004911.1 NC_004912.1 NC_004908.1 NC_004905.2 NC_004909.1 NC_004907.1 NC_004906.1
Influenza A (H2N2)	13,460	(−)ssRNA	NC_007378.1 NC_007375.1 NC_007376.1 NC_007374.1 NC_007381.1 NC_007382.1 NC_007377.1 NC_007380.1
Influenza A (H3N2)	13,630	(−)ssRNA	NC_007373.1 NC_007372.1 NC_007371.1 NC_007366.1 NC_007369.1 NC_007368.1 NC_007367.1 NC_007370.1
Influenza A (H1N1)	13,590	(−)ssRNA	NC_002023.1 NC_002021.1 NC_002022.1 NC_002017.1 NC_002019.1 NC_002018.1 NC_002016.1 NC_002020.1
Influenza A (H5N1)	13,590	(−)ssRNA	NC_007357.1 NC_007358.1 NC_007359.1 NC_007362.1 NC_007360.1 NC_007361.1 NC_007363.1 NC_007364.1
Influenza A (H7N9)	13,590	(−)ssRNA	KC885955 KC885956 KC885957 KC885958 KC885959 KC885960 KC885961 KC885962
Influenza B	14,450	(−)ssRNA	NC_002204.1 NC_002205.1 NC_002206.1 NC_002207.1 NC_002208.1 NC_002209.1 NC_002210.1 NC_002211.1
Parainfluenza 1	15,600	(−)ssRNA	NC_003461.1
Parainfluenza 2	15,650	(−)ssRNA	NC_003443.1
Parainfluenza 3	15,460	(−)ssRNA	NC_001796.2
Parainfluenza 4	17,050	(−)ssRNA	NC_021928.1
Human metapneumovirus	13,340	(−)ssRNA	NC_004148.2
Adenovirus C	35,937	dsDNA	NC_001405.1
Adenovirus B	35,343	dsDNA	NC_011203.1

(continued)

Table 1 (continued)

Respiratory virus (34)	Genome size	Type	Target genome sequences
Adenovirus E	35,994	dsDNA	NC_003266.2
Human coronavirus HKU1	29,930	(+)ssRNA	NC_006577.2
Human coronavirus NL63	27,550	(+)ssRNA	NC_005831.2
Human coronavirus 229E	27,320	(+)ssRNA	NC_002645.1
Human coronavirus OC43	30,738	(+)ssRNA	AY391777.1
Rhinovirus A	7150	(+)ssRNA	NC_001617.1
Rhinovirus C	7100	(+)ssRNA	NC_009996.1
Rhinovirus B14	7210	(+)ssRNA	NC_001490.1
Human Bocavirus 1	5299	ssDNA	NC_007455.1
Human Bocavirus 2	5196	ssDNA	NC_012042.1
Human Bocavirus 3	5242	ssDNA	NC_012564.1
Human Bocavirus 4	5104	ssDNA	NC_012729.2
KI polyomavirus	5040	dsDNA	NC_009238.1
WU polyomavirus	5229	dsDNA	NC_009539.1
HPeV type 1	7296	(+)ssRNA	FM242866.1
HPeV type 6	7347	(+)ssRNA	AB252582.1
Human enterovirus C104	7408	(+)ssRNA	AB686524.1
Human enterovirus C109	7354	(+)ssRNA	GQ865517.1

9. Microplate sealing film: Use one sheet to seal a 96-well microplate. Discard the seal after the removal from the plate. Use a new sheet to reseal the plate if needed.
10. Vortex mixer for small volumes in 96-well microplates and microcentrifuge tubes: Control speed and duration to obtain uniform mixing of solutions in microcentrifuge tubes or microplates.

11. Agilent 2100 Bioanalyzer, DNA 1000 (or 7500) chips and reagents. Store DNA chips at room temperature. Store the reagents at −20 °C. Before use equilibrate the reagents at room temperature for 30 min.
12. Magnetic beads separation stand for a 96-well microplate, magnetic beads separation rack for tubes.
13. Centrifuges: benchtop centrifuge with microplate carriers, microcentrifuge, mini centrifuge.
14. PCR Thermocycler for 96-well microplate.
15. Two water baths.
16. A set of pipettors with pipetting volumes from 2 μL to 1000 μL. One 20 μL 8-channel pipettor.

2.3 NGS Run Using MiSeq or NextSeq System

1. 0.5% Tween 20.
2. Illumina MiSeq Reagent kit v3 or v2 (*see* **Note 5**).
3. Molecular biology grade water.
4. Illumina MiSeq or NextSeq sequencing system.

3 Methods

Carry out all procedures at room temperature unless otherwise specified. Brief centrifugation or quick spin is done at 280 × *g* for 1 min in a benchtop centrifuge with microplate carriers for 96-well microplate or by a touch-spin using a microcentrifuge or a mini-centrifuge for microcentrifuge tubes. Thaw frozen reagents at room temperature completely, vortex and centrifuge briefly, then keep them at room temperature or on ice until use. Immediately return all reagents to their original storage condition after use except for Resuspension Buffer which is kept at 4 °C after the initial use.

3.1 TruSeq RVP RNA Access Library Preparation

3.1.1 Fragment RNA

1. Label a clean 96-well microplate as DFP, add 8.5 μL of Elute/Prime/Fragment High Mix and 8.5 μL of RNA samples (*see* **Note 6**) into each well, seal and shake the DFP plate at 1,600 rpm for 20 s.
2. Put the DFP plate in a thermocycler for incubation, using parameters: lid temperature at 100 °C, incubation at 94 °C for 8 min and then at 4 °C holding.
3. Remove the DFP plate from thermocycler when it reaches 4 °C, centrifuge, then proceed to **step 3.1.2** immediately.

3.1.2 Synthesize First Strand cDNA

1. Remove the seal from the DFP plate, add 7.2 μL of First Strand Synthesis Act D Mix and 0.8 μL of SuperScript III Reverse Transcriptase into each well, seal and shake the plate at 1,600 rpm for 20 s.

2. Put the DFP plate in the thermocycler for incubation, using parameters: lid temperature at 100 °C, incubation at 25 °C for 10 min, 42 °C for 15 min, 70 °C for 15 min, and finally at 4 °C holding.
3. Remove the DFP plate from the thermocycler when it reaches 4 °C, centrifuge, then proceed to **step 3.1.3** immediately.

3.1.3 Synthesize Second Strand cDNA

1. Remove the seal from the DFP plate, add 5 μL of Resuspension Buffer and 20 μL of Second Strand Marking Master Mix into each well, seal and shake the plate at 1,600 rpm for 20 s.
2. Put the DFP plate in the thermocycler for incubation, using parameters: lid temperature at 30 °C, incubation at 16 °C for 60 min and then at 24 °C for 2 min.
3. Immediately following the incubation, remove the DFP plate from thermocycler, remove the seal, add 90 μL of well-mixed AMPure XP beads to each well, seal and shake the plate at 1,800 rpm for 2 min.

   Follow **steps 4–10** in Subheading 3 for AMPure XP beads binding and washing.
4. Incubate the plate at room temperature for 5 min.
5. Centrifuge the plate briefly, place the plate on a magnet stand for 5 min.
6. Remove the seal from the plate, discard all supernatants from each well.
7. With the plate on the magnet stand, add 200 μL of 80% EtOH to each well.
8. Incubate the plate at room temperature for 30 s, then remove all supernatants.
9. Repeat **steps 7** and **8** once. Using a 20 μL multi-channel pipettor, carefully remove the remaining 80% EtOH from the wells.
10. Keep the plate stand at room temperature for 5 min to allow the EtOH to evaporate, then remove the plate from the magnet stand.
11. Add 17.5 μL of Resuspension Buffer into each well of the DFP plate, seal and shake the plate at 1,900 rpm for 2 min.
12. Incubate the DFP plate at room temperature for 2 min.
13. Centrifuge the DFP plate, and then place the plate on the magnet stand for 5 min.
14. With the DFP plate on the magnet stand, transfer 15 μL of clear supernatant from each well of the DFP plate to a new 96-well microplate labeled as ALP, without touching the magnetic beads. Seal the ALP plate. Discard the DFP plate.
15. Store the ALP plate at −20 °C and continue to **step 3.1.4** within 7 days.

3.1.4 Adenylate 3'End

If the ALP plate has been stored at −20 °C, thaw it at room temperature. Set up two water baths, one at 37 °C, the other at 70 °C.

1. Centrifuge the ALP plate. Remove the seal from the plate, add 2.5 μL of Resuspension Buffer and 12.5 μL of A-Tailing Mix to each well, seal the plate and shake it at 1,800 rpm for 2 min.
2. Incubate the ALP plate at 37 °C for 30 min, and then immediately transfer the plate to 70 °C and incubate for 5 min. After incubation, immediately place the plate on ice for at least 1 min.

3.1.5 Ligate Adapters

1. Remove the seal from the ALP plate, add 2.5 μL of Resuspension Buffer, 2.5 μL of Ligation Mix, and 2.5 μL of RNA Adapter Index to each well, seal and shake the plate at 1,800 rpm for 2 min, and then centrifuge the ALP plate.
2. Incubate the ALP plate at 30 °C for 10 min in a water bath, then move the plate to room temperature, and keep for 5 min to allow the ALP plate temperature drop to room temperature before proceeding to the next step.
3. Remove the seal from the ALP plate, add 5 μL of Stop Ligation Buffer to each well, seal and shake the plate at 1,800 rpm for 2 min, then centrifuge the plate.
4. Remove the seal from the ALP plate, add 42 μL of well-mixed AMPure XP beads to each well of the ALP plate, seal and shake the ALP plate at 1,800 rpm for 2 min.
5. Follow 3.1.3. **steps 4** to **10** for AMPure XP beads binding and washing.
6. Add 52.5 μL of Resuspension Buffer into each well of the ALP plate, seal and shake the ALP plate at 1,800 rpm for 2 min.
7. Incubate the ALP plate at room temperature for 2 min.
8. Centrifuge the ALP plate, remove the seal of the ALP plate, then place the plate on the magnet stand for 5 min.
9. Carefully transfer 50 μL of clear supernatant from each well of the ALP plate to a new 96-well microplate labeled as CAP. Discard the ALP plate.
10. Add 50 μL of well-mixed AMPure XP beads to each well of the CAP plate, seal and shake at 1,800 rpm for 2 min.
11. Follow 3.1.3. **steps 4** to **10** for AMPure XP beads binding and washing.
12. Add 22.5 μL of Resuspension Buffer into each well of the CAP plate, seal and shake the plate at 1,900 rpm for 2 min.
13. Incubate the CAP plate at room temperature for 2 min.
14. Centrifuge the CAP plate, remove the seal from the CAP plate, then place the plate on the magnet stand for 5 min.
15. Transfer 20 μL of clear supernatant from each well of the CAP plate to a new 96-well microplate labeled as PCR.

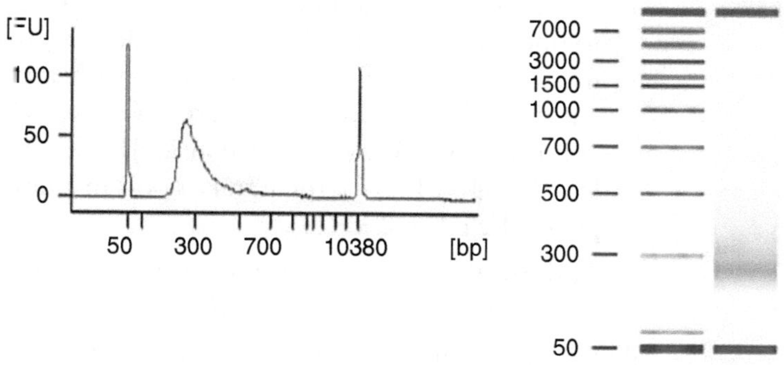

Fig. 2 Analysis of fragment library using Agilent 2100 Bioanalyzer and DNA 7500 kit

3.1.6 First PCR Amplification

1. Add 5 μL of PCR Primer Cocktail and 25 μL of PCR Master Mix to each well of the PCR plate.
2. Seal the PCR plate and shake at 1,600 rpm for 20 s. Centrifuge the PCR plate.
3. Put the PCR plate in a thermocycler for PCR amplification, using parameters: lid temperature at 100 °C, 98 °C for 30 s, then 15 cycles of 98 °C for 10 s, 60 °C for 30 s and 72 °C for 30 s, followed by 72 °C for 5 min and finally at 4 °C holding.
4. Remove the PCR plate from the thermocycler, centrifuge the plate, remove the seal, add 50 μL of well-mixed AMPure XP beads to each well, seal and shake the plate at 1,800 rpm for 2 min.
5. Follow 3.1.3. **steps 4** to **10** for AMPure XP beads binding and washing.
6. Add 17.5 μL of Resuspension Buffer into each well of the PCR plate, seal and shake the plate at 1,900 rpm for 2 min.
7. Incubate the PCR plate at room temperature for 2 min.
8. Centrifuge the PCR plate, remove the seal, then place the plate on the magnet stand for 5 min.
9. Transfer 15 μL of clear supernatant from each well of the PCR plate to a new 96-well microplate labeled as TSP1. Discard the PCR plate.

3.1.7 Quality Examination of the Library

Use 1 μL of each library from the TSP1 plate to measure DNA concentration and size distribution on Agilent 2100 Bioanalyzer with DNA 1000 (or 7500) kit (*see* **Note 7**). Determine DNA concentration in ng/μL for each library. The expected size distribution for the library is 200–500 bp with apparent peak at approximately 260 bp (Fig. 2).

3.1.8 First Hybridization

1. Combine up to four libraries (200 ng of DNA for each library) to one pool. If the total volume of a pool is greater than 45 μL, use a centrifugal ultrafiltration unit with a molecular weight cutoff of 30 kDa to concentrate the mixture. Bring the total volume to 45 μL with Resuspension Buffer if a pool volume is less than 45 μL. Save the library pools in a new 96-well microplate labeled as RAH1.
2. Add 50 μL of Capture Target Buffer 3 and 5 μL of the RVP oligos to each pool of the RAH1 plate.
3. Seal and shake the plate at 1,200 rpm for 1 min, then centrifuge the plate.
4. Put the plate in a thermocycler for incubation, using parameters: lid temperature at 100 °C, incubation at 95 °C for 10 min, 18 steps of decreasing temperatures starting at 94 °C, decreased by 2 °C at each step and kept for 1 min at each temperature, then final incubation at 58 °C for 90 min (*see* **Note 8**).
5. During the incubation, prepare Pre-Mixed Elution in a microcentrifuge tube (to be used in 3.1.9. **step** 7): mix 28.5 μL of Enrichment Elution Buffer 1 and 1.5 μL of 2 *N* NaOH for each pool, and keep at room temperature until use.
6. Once the incubation is over, immediately remove the plate from the thermocycler and proceed to **step 3.1.9**.

3.1.9 First Capture

1. Centrifuge the RAH1 plate, remove the seal, and transfer the first pool content to a clean 1.5 mL microcentrifuge tube labeled as S1, the second pool content to tube S2, and so on (*see* **Note 9**). Discard the RAH1 plate.
2. Add 250 μL of well-mixed Streptavidin magnetic beads to each S tube. Shake S tubes at 1,200 rpm for 5 min. Incubate at room temperature for 25 min. Centrifuge S tubes briefly, then put S tubes in a magnet rack for 2 min. Discard all supernatants with all S tubes on the magnet rack.
3. Remove all S tubes from the magnet rack, add 200 μL of Enrichment Wash Solution, shake at 1,800 rpm for 4 min.
4. Heat S tubes in a 50 °C water bath for 20 min. Then put S tubes on the magnet rack for 2 min. Remove supernatants with S tubes in the magnet rack.
5. Remove all S tubes from the magnet rack, add 200 μL of Enrichment Wash Solution, shake at 1,800 rpm for 4 min. Quick spin, then transfer the mixture from each S tube into a new 96-well microplate labeled as RAW1. Seal the RAW1 plate. Discard S tubes.
6. Incubate the RAW1 plate at the 50 °C water bath for 20 min. Immediately put the plate on the magnet stand for 2 min, then remove the supernatants with the plate on the magnet stand.
7. Add 23 μL of Pre-Mixed Elution to each well on the RAW1 plate, seal the plate and shake at 1,800 rpm for 2 min, incubate

the plate at room temperature for 2 min, centrifuge and place the plate on the magnet stand for 2 min.

8. Transfer 21 μL of the clear supernatant from each well of the RAW1 plate to a new 96-well microplate labeled as RAH2. Discard the RAW1 plate.
9. Add 4 μL of Elute Target Buffer 2 to each well of the RAH2 plate to neutralize the elution.
10. Seal the RAH2 plate, shake at 1,800 rpm for 1 min, and then centrifuge the plate.

3.1.10 Second Hybridization

1. Remove seal from the RAH2 plate, add 20 μL of Resuspension Buffer, 50 μL of Capture Target Buffer 3 and 5 μL of RVP oligos into each well.
2. Follow 3.1.8. **steps 3** to **6** for hybridization, then proceed immediately to **step 3.1.11**.

3.1.11 Second Capture

1. Centrifuge the RAH2 plate, remove the seal and transfer the contents from each well to clean 1.5 mL tubes labeled as S1, S2, and so on. Discard the RAH2 plate.
2. Follow 3.1.9. **steps of 2** to **10** for capture, except that the new microplates are labeled as RAW2 and RAC1 respectively.
3. Remove the seal from the RAC1 plate, add 45 μL of well-mixed AMPure XP beads to each well, seal and shake the plate at 1,800 rpm for 1 min.
4. Follow 3.1.3. **steps 4** to **10** for AMPure XP beads binding and washing.
5. Add 27.5 μL of Resuspension Buffer into each well, seal and shake the RAC1 plate at 1,900 rpm for 1 min.
6. Keep the RAC1 plate at room temperature for 2 min.
7. Centrifuge the RAC1 plate, remove the seal and place the plate on the magnet stand for 2 min.
8. Transfer 25 μL of the clear supernatant from each well of the RAC1 plate to a new 96-well microplate labeled as RAA. Seal the RAA plate. Discard the RAC1 plate.
9. Store the RAA plate at −20 °C and continue within a week.

3.1.12 Second PCR Amplification and Cleanup

1. Thaw the RAA plate, centrifuge the plate, remove the seal, add 5 μL of PCR Primer Cocktail and 20 μL of Enhanced PCR Mix. Seal the plate, shake at 1,200 rpm for 1 min, and then centrifuge the plate.
2. Put the RAA plate in thermocycler for PCR amplication, using parameters: lid temperature at 100 °C, 98 °C for 30 s, 17 cycles of 98 °C for 10 s, 60 °C for 30 s and 72 °C for 30 s, followed by 72 °C for 5 min and finally at 4 °C holding.

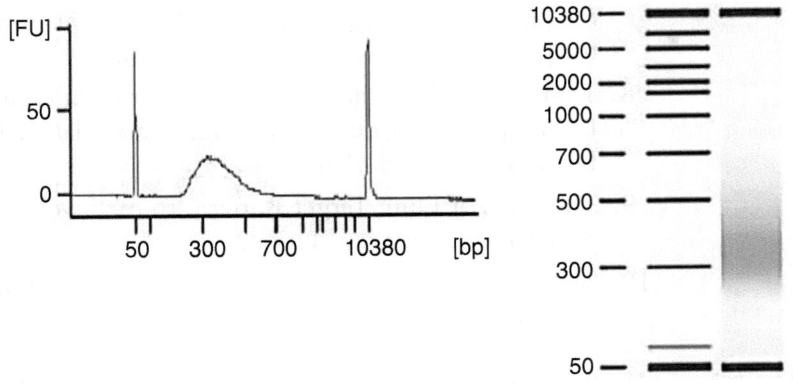

Fig. 3 Analysis of post-enrichment TruSeq RVP RNA Access library using Agilent 2100 Bioanalyzer and DNA 7500 kit

3. Remove the seal of the RAA plate, add 90 μL of well-mixed AMPure XP beads to each pool, seal and shake the plate at 1,800 rpm for 1 min.
4. Follow 3.1.3. **steps 4** to **10** for AMPure XP beads binding and washing.
5. Add 32 μL of Resuspension Buffer into each well of the RAA plate, seal and shake the plate at 1,900 rpm for 1 min.
6. Leave the RAA plate at room temperature for 2 min.
7. Centrifuge the RAA plate, remove the seal, and then place the plate on the magnet stand for 2 min.
8. Transfer 30 μL of clear supernatant from each well of the RAA plate to a new 96-well microplate labeled as RAL. Discard the RAA plate.

3.1.13 Quality Examination of the Finial Library

Use 1 μL of the final library in the RAL plate to measure DNA concentration and size distribution on Agilent 2100 Bioanalyzer with DNA 1000 (or 7500) kit. Determine DNA concentration in nM for each pool. The expected size distribution for the library is 200–500 bp with apparent peak at approximately 300 bp (Fig. 3) (*see* **Note 10**).

3.1.14 Dilution and Denaturation of the Libraries for NGS

The enriched libraries can be sequenced on Illumina Miseq or NextSeq systems. The loading concentration for libraries on MiSeq is 6–15 pM for reagents kit v2 or v3. We use loading concentration in the range of 8–14 pM for sequencing with MiSeq and reagents kit v3.

1. Dilute each library in the RAL plate from the concentration determined with Agilent 2100 Bioanalyzer to 1 nM with Resuspension Buffer. Prepare 22 μL of 1 nM of each library and pool all of them together into a new 1.5 mL tube labeled as 1 nM lib pool mix. Mix the tube at 1,200 rpm for 30 s and centrifuge. Keep the tube on ice until use. Store the RAL plate at −20 °C.

2. Prepare fresh 0.2 *N* NaOH by adding 200 μL of 1 *N* NaOH into 800 μL of nuclease-free water in a clean 1.5 mL tube labeled as 0.2 *N* NaOH. Mix the tube at 1,200 rpm for 30 s and centrifuge.
3. In a new 1.5 mL tube labeled as 20 pM lib pool mix, add 20 μL of 1 nM lib pool mix and 20 μL of 0.2 *N* NaOH. Mix the tube at 1,200 rpm for 30 s and centrifuge. Keep it at room temperature for 5 min. Add 20 μL of 0.2 M Tris-HCl, pH 7.0 into the tube. Mix the tube at 1,200 rpm for 30 s and centrifuge. Add 940 μL of prechilled HT1. Mix the tube at 1,200 rpm for 30 s and centrifuge. Keep the 20 pM lib pool mix tube on ice. Store the 1 nM lib pool mix tube at −20 °C.
4. In a new 1.5 mL tube labeled as loading conc. Pool, dilute the 20 pM denatured pool mix with prechilled HT1 using following chart as guide. Keep the loading conc. Pool tube on ice.

Loading concentration (pM)	6	8	10	12	14	20
20 pM lib pool mix (μL)	180	240	300	360	420	600
Pre-chilled HT1 (μL)	420	360	300	240	180	0

5. In a clean 1.5 mL tube labeled as 20 pM PhiX control (*see* **Note 11**), add 2 μL of 10 nM PhiX library, 3 μL of 10 nM Tris-Cl, pH 8.5 with 0.1% Tween 20, and 5 μL of 0.2 *N* NaOH. Mix the tube at 1,200 rpm for 30 s and centrifuge. Incubate at room temperature for 5 min. Add 990 μL of prechilled HT1. Mix the tube at 1,200 rpm for 30 s and centrifuge. Keep this tube on ice.
6. Add 6 μL of 20 pM PhiX control into 600 μL of the loading conc. pool tube. Mix the tube at 1,200 rpm for 30 s and centrifuge. Keep the tubes on ice until NGS run. (The mixture inside of the tube will be loaded into the reagent cartridge for sequencing run).

3.2 NGS and Sequence-Based Detection of Respiratory Pathogens

Sequence the lib pool with PhiX control using Illumina MiSeq or NextSeq system. The acquired sequence data can be applied to bioinformatics data analysis pipeline, which may include data quality processing, the removal of human genome and transcriptome sequences, de novo assembly, and comparison of sequence contigs and unassembled reads with sequence databases using BLAST programs [12] (*see* **Note 12**).

4 Notes

1. The method we described in this chapter is for research use only. It is not validated for clinical diagnosis purpose. The potential contamination and cross-contamination can come

from specimen collection, shipping and handling, storage, laboratory surfaces and instruments, reagents, viral cultures, nucleic acids or PCR amplification products, etc. Conduct the experiments in a clean laboratory room where high titer cultures, high concentration nucleic acids and PCR amplifications are not present. Clean frequently used instruments and devices, e.g., centrifuge, thermocycler, mixer, pipettors, on a regular basis to prevent accumulative contaminations and before each experiment to minimize carry-over of nucleic acids from the previous run. Use newly opened consumable materials and freshly prepared reagents. We highly recommend dividing original specimens, DNA/RNA extracts, and stock solutions into small aliquots or fractions to avoid accumulative contamination and for independent repeating of the test when trouble-shooting or verification of results is needed.

2. Respiratory swabs can be stored in transport medium such as Copan Universal Transport Medium (UTM) (Murrieta, CA, USA) or PrimeStore Molecular Transport Medium (MTM) (Longhorn Vaccines and Diagnostics LLC, Bethesda, MD, USA). We use QIAGEN QIAamp Viral RNA Mini Kit in DNA/RNA preparation. Carrier RNA is used in DNA/RNA extraction with QIAamp Viral RNA kit to increase RNA yield and protect purified RNA from degradation. Store RNA extracts in small aliquots and minimize freeze thaw cycles. Use RNase-decontamination products, for example, RNase AWAY made by Molecular Bio-Products, Inc., to clean the surface of the hood, benches, instruments, etc. When handle RNA samples, always wear gloves and change them often. During the procedure, work quickly and keep tubes closed and plates covered.

3. Most reagents used in this method are parts in Illumina TruSeq RNA Access Library Prep Kit. The user's guide that we used with modification is RS-301-9001DOC Part# 15049525 Rev. B. The plate names in the protocol are DFP (Depleted RNA Fragmentation Plate), ALP (Adaptor Ligation Plate), CAP (Clean up ALP Plate), RAH1 (RNA Access Hybridization 1), RAW1 (RNA Access Wash 1), RAH2 (RNA Access Hybridization 2), RAW2 (RNA Access Wash 2), RAC1 (RNA Access Clean-up 1), RAA (RNA Access Amplification), TSP1 (Target Sample Plate 1) and RAL (RNA Access Library).

4. SuperScript II Reverse Transcriptase and the new SuperScript IV Reverse Transcriptase can also be used. SuperScript III Reverse Transcriptase is the most often used reverse transcriptase. It is advised that SuperScript IV can obtain good yield and length of cDNA synthesis with improvements on thermal stability, speed, and resistance to inhibitors.

5. The TruSeq RNA Access library has size range around 300 bp. Both Illumina MiSeq Reagent kits v2 and v3, which can read

150 bp and 300 bp from either end, can be used in sequencing the libraries.

6. The TruSeq RNA Access Library guide recommends using 10–100 ng of RNA as starting materials. Since total nucleic acids extracted from respiratory specimens contain both DNA and RNA and have highly variable concentrations from specimen to specimen, we use 8.5 μL of DNA/RNA for each sample in RNA fragmentation.
7. Agilent 2100 Bioanalyzer can analyze up to 12 samples in one run. An alternative high-throughput system, the Agilent TapeStation 4200, can analyze up to 96 samples in one run.
8. The total hybridization time is about 2 h. Over hybridization may cause a high degree of nonspecific binding.
9. Instead of using microcentrifuge tubes and a magnetic beads separation rack, a 0.8 mL 96 deep well plate and a magnetic beads separation stand for 96-well plate can be used to process a large number of samples.
10. After two rounds of hybridization and capture and the second PCR amplification, the post-enrichment library appears having broadened size distribution with the peak shifted slightly to a higher molecular weight as compared with the fragment library shown in Fig. 2.
11. We always include Illumina PhiX Control v3 library in MiSeq runs. The PhiX control can be used for quick assessment of data quality and for technical trouble shooting. Besides, the addition of PhiX library can improve base calling quality for libraries with high G/C or A/T contents or low diversity sequences. The spiking ratio can be increased or reduced based on the needs and the nature of the specimens.
12. Using this method, partial to near-complete viral genome sequences can be obtained to allow genotyping and recombination and variant analysis. The method can detect both RNA and DNA viruses. During interpretation of the results, take cautious consideration of probable contamination and cross-contamination. If necessary, verify the results with additional tests such as molecular assays and/or by repeating the experiment with archived specimens or DNA/RNA aliquots.

Acknowledgments

We thank Darrell Dinwiddie for help in designing the RVP oligos. This work is supported by the Global Emerging Infections Surveillance and Response System (GEIS), a Division of the Armed Forces Health Surveillance Center. The views expressed here are those of the authors and do not reflect the official policy of the Department of the Army, Department of Defense or U.S. Government. We declare that no conflict of interest exists.

References

1. Garcia-Garcia G, Baux D, Faugere V et al (2016) Assessment of the latest NGS enrichment capture methods in clinical context. Sci Rep 6:20948
2. Terracciano I, Cantarella C, Fasano C et al (2017) Liquid-phase sequence capture and targeted re-sequencing revealed novel polymorphisms in tomato genes belonging to the MEP carotenoid pathway. Sci Rep 7(1):5616
3. Lee JS, Mackie RS, Harrison T et al (2017) Targeted enrichment for pathogen detection and characterization in three felid species. J Clin Microbiol 55(6):1658–1670
4. Deviatkin AA, Lukashev AN, Markelov MM et al (2017) Enrichment of viral nucleic acids by solution hybrid selection with genus specific oligonucleotides. Sci Rep 7(1):9752. https://doi.org/10.1038/s41598-017-10342-w
5. Bonfiglio S, Vanni I, Rossella V et al (2016) Performance comparison of two commercial human whole-exome capture systems on formalin-fixed paraffin-embedded lung adenocarcinoma samples. BMC Cancer 16:692
6. Wylie TN, Wylie KM, Herter BN et al (2015) Enhanced virome sequencing using targeted sequence capture. Genome Res 25(12):1910–1920
7. Briese T, Kapoor A, Mishra N et al (2015) Virome capture sequencing enables sensitive viral diagnosis and comprehensive virome analysis. MBio 6(5):e01491–e01415
8. Koehler JW, Hall AT, Rolfe PA et al (2014) Development and evaluation of a panel of filovirus sequence capture probes for pathogen detection by next-generation sequencing. PLoS One 9(9):e107007
9. Malhotra B, Swamy MA, Reddy PV et al (2016) Evaluation of custom multiplex real - time RT - PCR in comparison to fast - track diagnostics respiratory 21 pathogens kit for detection of multiple respiratory viruses. Virol J 13:91
10. Choudhary ML, Anand SP, Tikhe SA et al (2016) Comparison of the conventional multiplex RT-PCR, real time RT-PCR and Luminex xTAG(R) RVP fast assay for the detection of respiratory viruses. J Med Virol 88(1):51–57
11. Babady NE, Mead P, Stiles J et al (2012) Comparison of the Luminex xTAG RVP fast assay and the Idaho technology FilmArray RP assay for detection of respiratory viruses in pediatric patients at a cancer hospital. J Clin Microbiol 50(7):2282–2288
12. Kilianski A, Carcel P, Yao S et al (2015) Pathosphere.org: pathogen detection and characterization through a web-based, open source informatics platform. BMC Bioinformatics 16:416

Part IV

Oral Cavity

Chapter 11

Methods for Enrichment and Sequencing of Oral Viral Assemblages: Saliva, Oral Mucosa, and Dental Plaque Viromes

Marcos Parras-Moltó and Alberto López-Bueno

Abstract

The oral cavity is a major portal of entry for human pathogens including viruses. However, metagenomics has revealed that highly personalized and time-persistent bacteriophage assemblages dominate this habitat. Most oral bacteriophages follow lysogenic life cycles, deploying complex strategies to manage bacterial homeostasis. Although bacterial dysbiosis underlies common oral pathologies such as caries and periodontitis, the cause of these bacteria replacements remains obscure, and it is theorized that bacteriophages play an important role. The enormous sensitivity of metagenomics coupled with next-generation sequencing has made technically feasible to address the putative role of bacteriophages in oral dysbiosis and represents a valuable tool to discover new human viruses.

This chapter proposes a workflow that consists of a simple viral enrichment protocol, two alternative random amplification methods, and next-generation sequencing to access virome composition in three oral environments: supragingival plaque, saliva, and mucosa. These protocols circumvent some well-known sources of bias, providing genomic information about DNA and RNA viral communities with minimal contamination from human and bacterial sources.

Key words Oral viromes, Saliva, Oral mucosa, Supragingival plaque, Metagenomics

1 Introduction

The human oral cavity harbors an interesting variety of densely populated microbial ecosystems, including anaerobic niches covered by gingiva, soft tissues such as the tongue, soft palate or vestibular mucosa, the mineralized enamel surface of teeth, and saliva as the cleansing fluid that dilutes and mixes microbes from all oral environments. The mouth is one of the most diversely colonized sites in the human microbiome, with around 1000 indigenous bacterial species [1], a few archaea [2] and protozoan species [3], around 20 types of fungus [4], and 300–2000 viral genotypes [5] in any individual. These microorganisms are exposed to multiple

Andrés Moya and Vicente Pérez Brocal (eds.), *The Human Virome: Methods and Protocols*, Methods in Molecular Biology, vol. 1838, https://doi.org/10.1007/978-1-4939-8682-8_11,

and often changing, abiotic factors, including diverse oral surfaces that constrain bacterial adhesion, oral hygiene practices (tooth brushing and antiseptic mouthwash) that drastically reduce populations sizes, habits such as smoking, and continuous changes in osmotic conditions and pH promoted by intake of food containing a variety of chemicals [1, 6–8]. Despite the buffering action of saliva, these environmental factors make oral ecosystems highly dynamic. Other key factors that shape oral microbiota are the interaction with the host immune system tolerating symbiotic commensals, the frequent contact with environmental microorganisms through food intake or cohabiting, and the selective pressure exerted by viruses [9, 10].

Metagenomic studies have demonstrated that oral assemblages of viruses are dominated by bacteriophages [5, 11] that persist over the time in the mouth of healthy individuals [12, 13]. Although less abundant, eukaryotic DNA viruses including herpesvirus HPV7, several papillomaviruses [14], anelloviruses [15] as well as some RNA viruses [16] are also permanent constituents of a healthy mouth. Comparisons of oral viral assemblages among different individuals show gender-specific differences in composition [12] and remarkable interpersonal variability [5]. However, some viruses have been found ubiquitously among healthy individuals [12] and it is presumed that an increasing number of studies on oral viromes will permit the identification of "core" oral bacteriophages, in the same way as those found in gut samples [17, 18]. Interestingly, long-term antibiotic treatment notably modifies mouth viral assemblages [13, 19] and unrelated members of the same household share significant proportions of their saliva viromes, suggesting that environmental factors have a substantial impact on oral virus ecology [20].

Oral bacteriophages exert a plethora of poorly characterized host-control strategies [9]. Unlike the predominantly lytic interactions found in the predator-prey dynamic of aquatic bacteriophages, the human mouth is dominated by phages with lysogenic life cycles [5]. These temperate phages supply an extraordinary quantity of genes homologous to known virulence factors [5] that can modulate host phenotype. For instance, the *Streptococcus mitis* phage SM1 found in saliva encodes genes that mediate *S. mitis* attachment to platelets, which plays a major role in virulence of *S. mitis* in the endocardium [11].

The impact of the bacterial composition on oral health is well established, with complex consortia of bacteria associated with caries [3, 21], periodontitis [6, 22], and oral cancer [23]. Furthermore, multiple studies connect oral bacteria to systemic diseases [7]. In contrast, the role played by phages in host dysbiosis leading to oral pathology remains unexplored. Recently, significant expansion of myoviruses was detected in subgingival biofilms affected by periodontitis, suggesting the disease favors phages with lytic life cycles [24].

The human virome includes a wide range of viral size, morphology, and chemical composition that makes it unfeasible to set up universal protocols for viral metagenomics. Thus, currently available oral viromes have been obtained following purification and amplification steps that introduce bias to community composition. Commonly used 0.22 μm syringe filtration [5, 12, 13, 19, 20, 24] efficiently removes cellular contamination, but also large DNA viruses [11, 25] and reduces viral DNA yields to the half [26]. Similarly, relative phage abundances from oropharyngeal samples change drastically depending on whether samples were chloroform treated or 0.22 μm filtered in order to remove cellular contamination [11]. Ultracentrifugation through CsCl density gradients is another viral enrichment step extensively used in preparation of currently available oral viromes; this approach strongly discriminates against ssDNA viruses that sediments outside of the typically selected density layer [27–30] and affects viral stability of specific groups of phages [27, 29]. Finally, multiple displacement amplification (MDA) is required to obtain sufficient DNA template from oral viruses to utilize next-generation sequencing (NGS). This isothermal amplification is not exempt from bias as it preferentially amplifies circular DNA molecules over linear genomes [31] and introduces a systematic negative bias against CG-rich regions from nanograms of template [32–35].

Here, we present a simple viral enrichment protocol combined with two alternative random amplification methods, in order to provide comprehensive information about the human oral virome while avoiding some well-known sources of bias.

2 Materials

Solutions should be prepared in Milli-Q grade water (obtained by purifying deionized water, to attain a sensitivity of 18 MΩ-cm at 25 °C) or in commercial sterile nuclease-free distilled water, and preserved at room temperature (RT) unless otherwise indicated. Diligently follow all waste disposal regulations when disposing of waste materials.

In order to avoid contamination, the use of UV cabinet for the manipulation of DNA and RNA prior to random amplification is strongly recommended. Surfaces, micropipettes, and tip boxes should be carefully cleaned with a piece of paper rinsed in water with ionic detergent first, then acid-rinsed with 0.1 N HCl to break DNA molecules, and finally washed with Milli-Q water. Prior to use, expose consumable materials, micropipettes, and test tube racks to UV light for 1 h in the cabinet, replacing with new ones for every experiment.

2.1 Equipment

1. Micropipettes (P2, P20, P200, P1000).
2. UV cabinet for DNA/RNA manipulation.
3. Fume hood cabinet.
4. Vortex mixer.
5. General purpose benchtop centrifuge.
6. Ultracentrifuge and Beckman SW41 Ti rotor.
7. Thermoshaker.
8. PCR thermocycler.
9. UV transilluminator.
10. Fluorometer for PicoGreen® and RiboGreen® quantification.
11. Microvolume spectrophotometer.
12. Bioruptor Plus (Diagenode).
13. Illumina MiSeq or HiSeq sequencers.

2.2 Consumables

1. Catch-All™ Sample Collection Swabs (Epicentre).
2. Universal curette for supragingival sampling.
3. Sterile filter tips.
4. Protein low binding 1.5 mL tubes with Safe-Lock system, free of human DNA, DNase, RNase, and PCR inhibitors.
5. Conical centrifuge tubes of 15 and 50 mL.
6. PVDF syringe filters with 0.22 and 0.45 μm pore size, and 13 and 33 mm of diameter.
7. Glass Pasteur pipettes.
8. Transparent ultracentrifugation tubes, 17 mL.
9. 60% (w/v) Iodixanol (*OptiPrep*), Density Gradient Medium.
10. PCR tubes with flat cap (0.2 mL), RNase-/DNase-free, nonpyrogenic.
11. Surgical scalpel blades.

2.3 Reagents

1. Nuclease-free water: non DEPC-treated.
2. RT-PCR grade water: non DEPC-treated, eukaryotic and bacteria DNA free, DNA/RNA nucleases free.
3. 50,000 U/mL recombinant DNase I (grade I) in nuclease preservation buffer. Store aliquots at −70 °C for up to 18 months. Thaw only the amount needed for each experiment. Do not refreeze. Discard any leftover thawed solution.
4. 30,000 U/mL nuclease S7 in nuclease preservation buffer. Store aliquots at −70 °C up to 18 months. Thaw only the amount needed for each experiment. Do not refreeze. Discard any leftover thawed solution.
5. 20 mg/mL proteinase K in nuclease-free water. Stable up to 12 months at −20 °C.

6. 10 mg/mL RNase A in 10 mM Tris–HCl, pH 7.5, 15 mM NaCl. Store at −20 °C.
7. 10% SDS.
8. Phenol solution equilibrated in 10 mM Tris–HCl, pH 8.0 and 1 mM EDTA.
9. Phenol-chloroform solution (1:1). Mix vigorously with a vortex and allow phase separation in vertical position for at least 30 min.
10. Chloroform-isoamyl alcohol solution (24:1). Molecular biology grade isoamyl alcohol. Mix vigorously with a vortex.
11. TRIzol-LS® Reagent.
12. Mussel glycogen. Prepare at 20 µg/µL in RT-PCR grade water (*see* **Note 1**).
13. Absolute ethanol. Ethanol can be diluted to 70 and 75% in nuclease-free water.
14. Absolute isopropanol.
15. RNase-free DNase.
16. Illustra™ Ready-to-go™ GenomiPhi™ V3 DNA Amplification kit (GE Healthcare).
17. TruePrime™ WGA Kit (Sygnis Biotech).
18. SuperScript™ III First-Strand Synthesis System for RT-PCR (Invitrogen).
19. Klenow fragment (3′-5′ exo-).
20. Q5 Hot Start high fidelity DNA polymerase (NEBiolabs).
21. Agarose D1 Low EEO and high gel strength.
22. Gel Extraction Kit.
23. Pseudodegenerated primers: FR26RV-12 N 5′-GCCGGAGCTCTGCAGATATCNNNNNNNNNNNN-3′), 454-A-12 N (5′- ATCGTCGTCGTAGGCT GCTNNNNNNNNNNNN-3′), PrimerK-12 N (5′-GACCATCTAGCGACCTCCACNNNNNNNNNNNN-3′).
24. Universal primers: FR20RV 5′-GCCGGAGCTCTGC AGATATC-3′),454-A(5′-ATCGTCGTCGTAGGCTGCT-3′), PrimerK (5′- GACCATCTAGCGACCTCCAC-3′).

2.4 Buffers

All buffers should be prepared inside the UV cabinet.

1. 1 M Tris–HCl pH 7.5: Dissolve 6.05 g of Tris base in 30 mL of water. Adjust pH to 7.5 with 5 M HCl. Adjust final volume to 50 mL with Milli-Q water [36].
2. Saline Magnesium buffer (SM buffer10×): 500 mM Tris–HCl pH 7.5, 1 M NaCl, 100 mM SO_4Mg. Mg^{2+} is required to preserved phage tail integrity. Prepare in nuclease-free water, dispense into aliquots after filter-sterilization and preserve frozen.

3. Nuclease preservation buffer: 20 mM Tris–HCl pH 7.5, 1 mM $MgCl_2$. Prepare in nuclease-free water, dispense into aliquots after filter-sterilization and preserve frozen.
4. Nuclease reaction buffer 20×: 200 mM Tris–HCl pH 7.5, 200 mM $MgCl_2$, 40 mM $CaCl_2$. Prepare in nuclease-free water, dispense into aliquots after filter-sterilization, and preserve frozen.
5. 0.5 M EDTA: Add 186.1 g of disodium EDTA to 800 mL with Milli-Q water; stir vigorously on a magnetic stirrer. Adjust the pH to 8.0 with NaOH (~20 g of NaOH pellets) and dispense into aliquots and sterilize by autoclaving [36].
6. 0.5 M EGTA: Add 19.01 g ethylene glycol bis (b-aminoethyl ether) N,N,N′,N′-tetraacetic acid (EGTA; FW 380.35) to 80 mL with Milli-Q water. Stir vigorously, heating slightly. Adjust pH to 8.0 with NaOH to permit EGTA to go into solution. Adjust final volume to 0.1 L with Milli-Q water, store in metal-free plastic aliquots, and sterilize by autoclaving [36].
7. 3 M NaAc pH 5.6: Dissolve 246.1 g of sodium acetate in 500 mL of deionized Milli-Q water. Adjust the pH to 5.6 with glacial acetic acid. Allow the solution to cool overnight. Adjust the pH once again to 5.6 with glacial acetic acid and the final volume to 1 L with Milli-Q water. Dispense into aliquots and sterilize by autoclaving [36].

3 Methods

The next workflow combines three slightly different sampling methods for three oral ecosystems with a simple viral enrichment protocol based on centrifugation, filtration, and an optional iodixanol cushion. Subsequent nuclease treatment digests non-capsid protected genetic material before DNA extraction with proteinase K-SDS and phenol or RNA extraction with TRIzol-LS®. Two random amplification methods, MDA and sequence independent single primer amplification (SISPA), provide enough DNA material to perform NGS with Illumina systems (Fig. 1).

We have set up these protocols with pooled saliva, oral mucosa, and dental plaque samples. Viromes obtained by this method exhibit similar proportions of bacteria, viral and unknown sequences, but differ in composition of viral families. Interestingly, comparison of MDA-amplified and unamplified saliva viromes shows a nearly identical profile of phage families, suggesting that MDA does not introduce significant bias at this taxonomic level (Fig. 2).

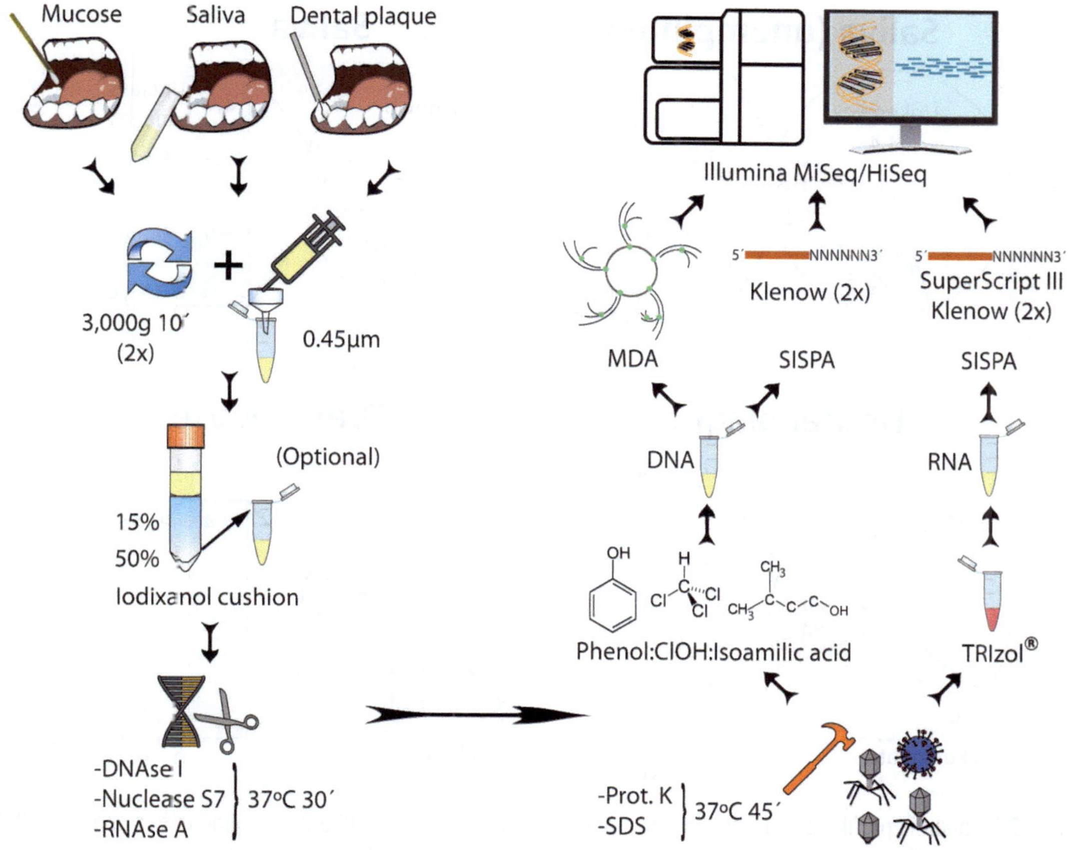

Fig. 1 Workflow diagram for metagenomic analysis of viruses from three oral environments: mucosa, saliva, and supragingival plaque. This protocol includes sampling, viral enrichment (centrifugation and filtration), nuclease treatment, DNA/RNA extraction (Phenol and *Trizol-LS*® reagent™), random amplification (MDA and SISPA), and Illumina sequencing (*MiSeq* and *HiSeq*)

3.1 Sampling

Oral samples should be collected at least 3 h after last food intake or oral-hygiene measures such as tooth brushing or antiseptic mouthwashes. This lapse of time reduces contamination from food-associated viruses and increases biological material collection. Negative controls should be carried out in parallel across the whole procedure including Milli-Q water sampling with sterile curettes or swabs, as well as treatments with the same consumables, buffers, and reagents used for each sample.

3.2 Sampling Saliva

1. Let saliva collect in the mouth for at least 1 min.
2. Drool 3 mL of non-stimulated saliva into sterile 15 mL tubes (*see* **Note 2**).
3. Dilute samples with 3 volumes of 1× SM buffer to reduce saliva viscosity (*see* **Note 3**) and avoid filter clogging.

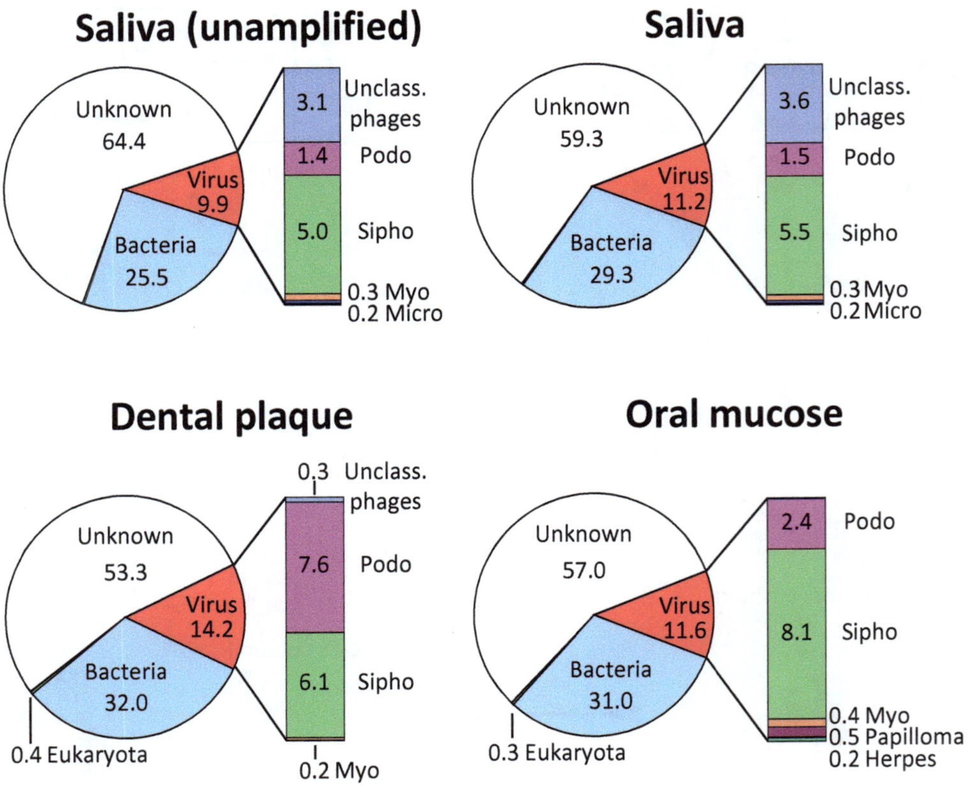

Fig. 2 Taxonomic profiles of four oral viromes obtained following the workflow described in this chapter (MDA amplification). Subsamples of 10,000 reads from each virome were BLASTx compared to the GenBank non-redundant protein database (nr); best hits with e-value <0.001 are shown in percentage at both kingdom and viral family level. Note that community composition at these taxonomic levels was similar before and after MDA amplification for the saliva sample. On the other hand, there are notable differences among viral family profiles of saliva, dental plaque, and oral mucosa

4. Vigorously shake by vortex until no visible clumps are observed. Combine vortex with ice-cooling to prevent overheating.
5. Place the sample on ice for not more than 4 h and proceed with the virus enrichment protocol or preserve at −80 °C.

3.3 Sampling Oral Mucosa

1. Swab the entire area of both left and right buccal mucosa and the inferior vestibule for 10 s each. Take care not to touch the teeth [37].
2. Swirl the swab sponge in 0.8 mL of 1× SM buffer for 30 s and press against the tube bottom multiple times to ensure transfer of bacteria from swab to solution.
3. Place the sample on ice and proceed with the viral enrichment protocol within the next 4 h. Otherwise, the sample should be preserved at −80 °C with 1× SM supplemented with 20% sucrose to mitigate capsid or envelope disruption during freezing and thawing.

3.4 Sampling Supragingival Plaque

1. Retrieve supragingival plaque with a universal curette from each tooth surface with as many strokes as necessary.
2. Immerse the curette tip in 0.5 mL 1× SM buffer and shake vigorously for 5 s.
3. Sample the same site again to collect as much material as possible and pool together plaque material from all tooth surfaces if necessary.
4. Vortex vigorously for 30 s to maximize dispersion of biological material in the fluid. Put the tube on ice immediately to prevent overheating. Repeat twice.
5. Place the sample on ice and proceed with the viral enrichment protocol within the next 4 h. Otherwise, the sample should be preserved at −80 °C in 1× SM buffer supplemented with 20% sucrose to mitigate capsid or envelope disruption during freezing and thawing.

3.5 Centrifugation and Filtration

1. Add 1 volume of 1× SM buffer to mucosa and dental plaque samples to reduce the density of the aqueous phase during the subsequent phenol extraction.
2. Homogenize samples by vortexing vigorously for 30 s immediately followed by cooling tubes on ice for 30 s to prevent temperature increase. Repeat this step twice.
3. Centrifuge for 10 min at 3000 × *g* at RT to reduce cellular contamination. Transfer the supernatant to a fresh 1.5 mL tube and repeat the centrifugation (*see* **Note 4**).
4. Filtrate the supernatant through 0.45 μm syringe filters. Use 13 mm diameter filters and sterile 1 mL syringes for mucosa and dental plaque samples, and 33 mm diameter filters and sterile 20 mL syringes for saliva samples. Replace filters with new ones as necessary to avoid clogging (*see* **Note 5**).
5. Collect filtered samples in 1.5 mL tubes (mucosa and dental plaque) and 15 mL tubes (saliva). Use a P200 micropipette to recover filtered dead volume at the bottom part of the filter.

3.6 Iodixanol Cushion (See Note 6)

1. Dilute filtered sample with 1× SM buffer to reach a final volume of 12 mL and pour the diluted sample into a transparent 17 mL ultracentrifugation tube.
2. Clean glass Pasteur pipettes by plunging into 0.1 N HCl for 10 min and wash extensively with Milli-Q water. Then, introduce a cleaned Pasteur pipette into the tube.
3. Deposit 3 mL of 15% (w/v) iodixanol prepared in 1× SM buffer at the bottom of the tube by applying a continuous flow (use P1000 micropipette) through the Pasteur pipette.
4. Add 0.5 mL of 50% (w/v) iodixanol solution prepared in 1× SM buffer at the bottom of the tube in a similar way. Avoid

flow interruption while adding iodixanol solutions to prevent bubbles that might disturb the cushion layers.

5. Carefully remove the Pasteur pipette and equilibrate tubes by slowly adding 1× SM buffer to the top of the tube.
6. Centrifuge at 18,000 × *g* for 16 h in an ultracentrifuge using a Beckman SW41Ti rotor, with slow brake speed options.
7. Remove liquid from the top with a P1000 micropipette and collect around 1 mL containing viral particles from the interphase between the 15/50% iodixanol layers (*see* **Note 7**).

3.7 Nuclease Treatment

1. Mix the recovered fraction from the iodixanol cushion with 1 volume of nuclease-free water in 1.5 mL safe-locker tubes. This step is intended to dilute iodixanol concentration during phenol and TRIzol-LS® extractions.
2. Add 20× nuclease buffer to reach a final concentration of 1×.
3. Treat samples with a nuclease cocktail for 30 min at 37 °C, with agitation. This cocktail is composed of 250 U/mL DNase I, 250 U/mL nuclease S7 and 100 μg/mL RNase A to remove DNA/RNA molecules not protected by capsid or viral envelopes (*see* **Note 8**).
4. Stop nuclease activity by adding 20 nM EDTA and 2 nM EGTA. Mix vigorously and incubate for 5 min at RT.
5. Disrupt capsid and viral envelopes with 200 μg/μL of proteinase K and 0.5% SDS for 45 min at 37 °C.

3.8 Phenol Extraction of Viral DNA Genomes (See Note 9)

1. Add 1 volume of phenol equilibrated in 10 mM Tris–HCl pH 7.5 and vortex vigorously for 30 s. Centrifuge at 8000 × *g* for 4 min at RT and carefully collect the aqueous phase with a P1000 without touching the tube walls, adding it to a fresh 1.5 mL tube. Do not disturb the interphase between the aqueous phase (top layer) and the organic phase (bottom layer). Precaution: work in a fume hood and wear gloves and goggles.
2. Repeat the same procedure with 1 volume of phenol:chloroform (1:1) solution. Precaution: work in a fume hood and wear gloves and goggles.
3. Repeat the same procedure with 1 volume of chloroform:isoamyl alcohol (24:1) solution. Precaution: work in a fume hood and wear gloves and goggles.
4. Add 20 μg of mussel glycogen as a carrier for DNA precipitation and mix by pulse-vortexing for 5 s.
5. Add 0.1 volume of 3 M NaAc pH 5.6.
6. Add 2.5 volumes of absolute ethanol and mix by pulse-vortexing for 5 s.

7. Let the DNA precipitate overnight at −80 °C.
8. Centrifuge at 12,000 × *g* for 30 min at 4 °C. Discard the supernatant.
9. Add 1 mL of cold 70% ethanol.
10. Centrifuge at 12,000 × *g* for 10 min at 4 °C. Discard the supernatant. Repeat this step again.
11. Air-dry the pellets at RT in open tubes for 5 min.
12. Resuspend in 20 μL of 15 mM Tris–HCl pH 7.5 prepared in RT-PCR grade water by incubating at 50 °C for 5 min and pipetting up and down several times with a P20.

3.9 TRIzol-LS® Extraction of Viral RNA Genomes

RNA viral genomes are purified with TRIzol-LS Reagent using the supplier's instructions with the following minor modifications:

1. Add 3 volumes of TRIzol-LS® to 0.3 mL of nuclease-treated sample. The purified viral sample is considered to be a "biological fluid."
2. Skip the optional centrifugation step before phase separation.
3. Incubate for 15 min at RT after adding chloroform.
4. Avoid dragging any material from the interphase by carefully pipetting with a P200 and maintaining at least 3 mm of separation from both the interphase and the tube wall.
5. Add 20 μg of nuclease-treated mussel glycogen as carrier.
6. Let the RNA pellet air-dry for a maximum of 5 min.
7. Resuspend the pellet in 34 μL of RT-PCR grade water and 1 μL of RNaseOUT™ by pipetting up and down several times with a P20, and incubate in a thermoshaker 15 min at 60 °C (*see* **Note 10**).
8. Treat with RNase-free DNase following the supplier's instructions.
9. Precipitate again with glycogen as described above and resuspend in 12 μL of RT-PCR grade water with 1 μL of RNaseOUT™.

3.10 Quantification of Extracted Viral Genomes

The concentration of purified viral DNA/RNA must be quantified using PicoGreen® or RiboGreen® technologies since typical concentration of material recovered following this protocol is less than 1 ng/μL which is below the limit of detection of microvolume spectrophotometer.

3.11 Multiple Displacement Amplification

Illustra Ready-To-Go GenomiPhi™ V2 and V3 DNA Amplification Kits and TruePrime™ WGA Kit are two procedures based on phi29 phage polymerase isothermal amplification that differ in their random priming strategy: the former uses random hexamers and the latter uses random short DNA primers synthetized in the reaction tube by the primase activity of *Tth*PrimPol. Primase-based strategy reduces background amplification in the absence of input DNA, improves coverage evenness [38], and provides better bacterial community representation [39] (*see* **Note 11**).

1. For GenomiPhi™ amplification, use the manufacturer's instructions with the following minor modifications:
   (a) Utilize >1 ng of viral DNA diluted in RT-PCR grade water as template. The less template used in the reaction, the more skewed the viral community will be after amplification.
   (b) Use temperature-based denaturation instead of the chemical alternative (*see* **Note 12**).
   (c) After denaturation and subsequent cooling, carefully transfer all liquid and drops from the tube wall to the tubes with lyophilized enzyme.
   (d) Incubate for not more than 2.5 h at 30 °C. Run a negative control in parallel for every experiment to check for background amplification.
   (e) Precipitate MDA products by adding 150 μL of absolute ethanol at RT for 15 min to reduce salts, nucleotides, and random primers.
   (f) Centrifuge at 12,000 × *g* for 2 min at RT.
   (g) Wash with 100 μL of 70% ethanol.
   (h) Centrifuge at 12,000 × *g* for 2 min at RT.
   (i) Resuspend in 50 μL of 15 mM Tris–HCl pH 7.5.
   (j) Pipette the MDA-product up and down multiple times with a P20 and incubate for 10 min at 50 °C (*see* **Note 13**).
   (k) Store amplified DNA at 4 °C for short-term storage or at −20 °C for long-term storage. After thawing, incubate again for 10 min at 50 °C.
2. For TruePrime™ amplification, follow the supplier's instructions with the next minor modifications:
   (a) Utilize >1 ng of viral DNA diluted in RT-PCR grade water as template in the reaction. The less template used in the reaction, the more skewed the viral community will be after amplification.
   (b) Incubate at 30 °C for 3 h (*see* **Note 14**).
   (c) Run a negative control in parallel for every experiment to check for background amplification.
   (d) Precipitate and store TruePrime™ product as described above for GenomiPhi™-products.

3.12 Sequence Independent Single Primer Amplification (SISPA) for RNA Genomes

SISPA is a PCR-based approach that uses pseudodegenerated oligonucleotides. We describe a protocol based on previous reports [40–43] with some modifications. The random region at the 3′ end of the pseudodegenerated primer is extended to 12 Ns to improve random annealing [44].

1. Subject purified viral genomes to reverse transcription following SuperScript III manufacturer's instructions specified for random priming.
2. In three separate 0.2 mL tubes, combine 5 μL of purified RNA, 1 μL of dNTPs mix (10 mM of each), and 60 pmol of one of the following pseudodegenerated primers: FR26RV-12 N, 454-A-12 N or PrimerK-12 N. Complete with RT-PCR grade water to reach a final volume of 10 μL.
3. Incubate at 65 °C 5 min for denaturation.
4. Cool down sample at 4 °C for 2 min to allow primer annealing.
5. Add prepared RT reaction mix (2 μL 10× buffer, 4 μL 25 mM $MgCl_2$, 2 μL 0.1 M DTT, 1 μL RNaseOUT™ and 1 μL SuperScript III enzyme).
6. Incubate in a thermocycler for 10 min at 25 °C, 60 min at 50 °C (RNA turns into ssDNA), 5 min at 80 °C (polymerase inactivation) and cool to 4 °C.
7. Add 0.5 μL of RNase H and heat to 37 °C for 10 min to disrupt DNA/RNA hybrids.
8. Heat to 94 °C for 3 min (DNA denaturation).
9. Put the sample immediately on ice and add 1.5 μL of 10× Klenow buffer, 12 μL of RT-PCR grade water, and 1 μL of Klenow enzyme.
10. Perform Klenow elongation at 37 °C for 1 h (dsDNA generated from ssDNA template).
11. Inactivate the polymerase by heating to 75 °C for 10 min.
12. PCR amplification with universal primers. Combine 5 μL of dsDNA template with 10 μL of 5× Q5 polymerase buffer, 1 μL of 25 mM $MgCl_2$, 1.5 μL of 10 mM dNTPs mix, 0.7 μL of Q5 polymerase, and 4 μL of the corresponding universal primer at 0.8 mM: FR20RV, 454-A, or PrimerK. Complete with RT-PCR grade water to a total volume of 50 μL (*see* **Note 15**).
13. In a thermocycler, follow this temperature protocol: denaturation at 98 °C for 2 min + 35 cycles (10 s at 98 °C, 30 s at 65 °C and 1 min 15 s at 72 °C), and final elongation at 72 °C for 2 min 30 s.
14. Analyze the amplification by electrophoresis in 1% agarose gel, loading 5 μL of the PCR product to one well for analytical purposes and the rest of the volume for gel-band extraction.

15. Cut the DNA smear between 0.8 and 1.5 kb with a scalpel. Use a fresh scalpel for each gel-band to avoid cross-contamination.
16. Weigh the sliced gel and place into a fresh 15 mL tube (*see* **Note 16**).
17. Purify sliced gel-bands using a gel extraction kit following the manufacturer's instructions. Perform all optional purification steps for more efficient sequencing.
18. Elute the sample in 40 μL nuclease-free water.

3.13 SISPA for DNA Genomes

1. Convert ssDNA to dsDNA molecules by two rounds of Klenow extension: combine >1 ng of extracted DNA with 2 μL of 10× Klenow buffer, 1 μL of dNTPs mix (10 mM each), 60 pmol of one of the following pseudodegenerated primers: FR26RV-12 N, 454-A-12 N, or PrimerK-12 N and 1 μL of Klenow enzyme. Complete with RT-PCR grade water to reach a final volume of 20 μL.
2. Perform Klenow elongation at 37 °C for 1 h (ssDNA become dsDNA) from ssDNA.
3. Denature DNA by heating to 94 °C for 3 min.
4. Add 1 μL of Klenow enzyme and repeat **step 2**.
5. Follow the protocol described above (Subheading 3.12), starting at **step 11** (*see* **Note 17**).

3.14 NGS Sequencing

Excellent reviews compare the advantages and disadvantages of available NGS systems [45, 46]. Among them, Illumina sequencers provide millions of high-quality reads, and are currently the technique of choice in most metagenomics studies. There are several convenient methods for library preparation that are compatible with Illumina sequencers such as Nextera®, TruSeq® and NEBNext® Ultra™ DNA Library Preparation kits for Illumina®. We do not have experience with Nextera® libraries, but some studies conclude that caution should be taken with such transposon-based library preparation procedures since they produce more biased viral assemblages than TruSeq® [47].

In our workflow, NEBNext® libraries are prepared with >0.5 μg of random amplified DNA fragmented to average lengths of 700–1000 nt by sonication. By using these libraries, we obtain 1–2 millions of 2 × 250 bp or 2 × 300 bp pair-end reads in MiSeq sequencers (MiSeq Reagent Kits v2: 500 cycles or MiSeq Reagent Kit v3: 600 cycles) and 10–20 millions of 2 × 250 bp reads in HiSeq 2500 sequencers (HiSeq Rapid SBS Kit v2: 500 cycles). Rapid advances in NGS techniques together with falling prices will continue to increase read length and output by the time this chapter is printed. Therefore, a search for updated comparisons of NGS technologies is highly recommended.

4 Notes

1. Although this carrier is claimed to be free of nucleic acids, nucleases, and proteases, some authors have detected trace amounts of genetic material in glycogen stocks from two out of nine anonymous suppliers [48]. Therefore, we heat glycogen at 95 °C for 15 min and treat with nucleases following the protocol described below for viral genome purification (*see* Subheadings 3.7 and 3.8).
2. Alternatively, saliva production can be stimulated by administration of a standard piece of unflavored gum base (1.0–1.5 g; Wrigley Co., Peoria, IL). This process may be repeated multiple times in order to collect large volumes of saliva.
3. Some authors recommend incubating the sample with dithiothreitol (DTT) 2.17 mM for 30 min at 37 °C to break up mucous [49]. However, DTT cleaves disulfide linkages among cysteines, disrupting the envelope and coat of herpesviruses, poxviruses, paramyxoviruses, and polyomaviruses, among others [50].
4. Combine pellets in 1× SM buffer if analysis of cellular genomes is desired.
5. In our hands, two consecutive centrifugation steps helped to reduce >10^4 times the number of colonies of *Escherichia coli* and >10^2 times the number of colonies of small-size bacteria such as *Staphylococcus aureus* and *Roseobacter litoralis* (unpublished results). Further combination with 0.45 μm filtration decreased the number of colonies >10^7 times in all cases. However, oral viromes prepared following our workflow showed contamination with reads from small bacteria related to candidate division TM7 (unpublished results). Alternative 0.22 μm filtration enhances removal of small bacteria but also of large viruses [11, 25] and introduces an additional source of bias, as a significant reduction of viral DNA yields has been reported in comparison to 0.45 μm filters [26].
6. Iodixanol cushion reduces the amount of low-density material including contaminating free DNA. However, this step can be skipped for mucosa and dental plaque without a significant impact in community composition. We strongly recommend iodixanol cushion for concentration of saliva samples since direct concentration by ultracentrifugation might affect virus stability. 15/50% iodixanol solutions (w/v) can be replaced by 20/70% (w/v) sucrose.
7. Keep a small aliquot of 10–50 μL after virus enrichment to estimate cellular contamination by EM [51], epifluorescence [52], and PCR against bacterial 16S rRNA genes [53].
8. An effective nuclease treatment is a crucial step in this protocol. Other authors have used *Turbo DNase* (Ambion) and/or *Benzonase* (Novagen) together with RNases (One or A) [40].

9. It is possible to extract DNA using commercial purification kits. However, these kits do not outperform phenol extraction in terms of DNA yield [54], fail to purify efficiently large DNA molecules which are partially retained in the column silica matrix during elution, and can be an unexpected source of contamination [55, 56].
10. The RNA is often invisible prior to 75% ethanol centrifugation, and forms a gel-like pellet on the side and bottom of the tube.
11. Linker Amplification Shotgun Libraries (LASL) is an alternative random amplification protocol frequently used in metagenomic analysis of viral communities [57]. This PCR-based protocol systematically under-amplifies regions with high CG %, and although it traditionally targets only dsDNA, a recent modification of the protocol extends LASL analysis to ssDNA genomes [58]. However, LASL requires previous *Covaris* fragmentation and DNA-size selection that makes this approach useless when DNA amount is a limiting factor (as is the case in oral samples) [59].
12. Over-representation of circular single-stranded genomes can be deliberately exacerbated by skipping the denaturation hybridization step [31].
13. MDA product is a complex structure of >10 kb molecules that gives variable measures during quantification. The heating step likely reduces the complexity of this structure and allows for better incorporation of *PicoGreen*® reagents.
14. Incubation time can be increased to 6 h if higher amplification yields are needed.
15. Using equimolar mixtures of each modified universal primer with 0–4 N's at the 5′ end during PCR amplification improves the identification of clusters during Illumina sequencing [60].
16. Weigh the tube with and without the gel-band to determine mass of gel-band.
17. Pooling PCR products obtained with three different primers reduced coverage bias due to the preferential annealing to sequences with high identity [44].

Acknowledgments

This work was supported by the Spanish Ministry of Economy and Competitiveness through grant SAF2012-38421 and a "Formación de Personal Investigador" Ph.D. studentship to M. P-M.

We acknowledge Patricia Suárez Rodríguez and Ana Rodríguez Galet for technical support and Áurea Soro-Simón, Alex Mira, Asier Eguía and José Manuel Aguirre-Urizar for providing valuable samples.

References

1. Dewhirst FE, Chen T, Izard J et al (2010) The human oral microbiome. J Bacteriol 192(19):5002–5017
2. Matarazzo F, Ribeiro AC, Feres M et al (2011) Diversity and quantitative analysis of Archaea in aggressive periodontitis and periodontally healthy subjects. J Clin Periodontol 38(7):621–627
3. Belda-ferre P, Alcaraz LD, Pignatelli M et al (2012) The oral metagenome in health and disease. ISME J 6(1):46–56
4. Ghannoum MA, Jurevic RJ, Mukherjee PK et al (2010) Characterization of the oral fungal microbiome (mycobiome) in healthy individuals. PLoS Pathog 6(1):e1000713
5. Pride DT, Salzman J, Haynes M et al (2012) Evidence of a robust resident bacteriophage population revealed through analysis of the human salivary virome. ISME J 6(5):915–926
6. Camelo-Castillo AJ, Mira A, Pico A et al (2015) Subgingival microbiota in health compared to periodontitis and the influence of smoking. Front Microbiol 6:1–12
7. Wade WG (2013) The oral microbiome in health and disease. Pharm Res 69(1):137–143
8. Proctor DM, Relman DA (2017) The landscape ecology and microbiota of the human nose, mouth, and throat. Cell Host Microbe 21(4):421–432
9. Edlund A, Santiago-Rodriguez TM, Boehm TK et al (2015) Bacteriophage and their potential roles in the human oral cavity. J Oral Microbiol 7:27423
10. Belkaid Y, Hand TW (2014) Role of the microbiota in immunity and inflammation. Cell 157(1):121–141
11. Willner D, Furla M, Schmieder R et al (2011) Metagenomic detection of phage-encoded platelet-binding factors in the human oral cavity. Proc Nat Acad Sci 108:4547–4553
12. Abeles SR, Robles-Sikisaka R, Ly M et al (2014) Human oral viruses are personal, persistent and gender-consistent. ISME J 8(9):1753–1767
13. Ly M, Jones MB, Abeles SR et al (2016) Transmission of viruses via our microbiomes. Microbiome 4(1):64
14. Wylie KM, Mihindukulasuriya K, Zhou Y et al (2014) Metagenomic analysis of double-stranded DNA viruses in healthy adults. BMC Biol 12:71
15. Parras-Moltó M, Suárez-Rodríguez P, Eguia A et al (2014) Genome sequence of two novel species of torque Teno Minivirus from the human oral cavity. Genome Announc 2(5):5–6
16. Corstjens PLAM, Abrams WR, Malamud D (2016) Saliva and viral infections. Periodontol 2000 70(1):93–110
17. Stern A, Mick E, Tirosh I et al (2012) CRISPR targeting reveals a reservoir of common phages associated with the human gut microbiome. Genome Res 22(10):1985–1994
18. Manrique P, Bolduc B, Walk ST et al (2016) Healthy human gut phageome. Proc Nat Acad Sci 113(37):10400–10405
19. Robles-Sikisaka R, Ly M, Boehm T et al (2013) Association between living environment and human oral viral ecology. ISME J 7(9):1710–1724
20. Abeles SR, Ly M, Santiago-Rodriguez TM et al (2015) Effects of long term antibiotic therapy on human oral and fecal viromes. PLoS One 10(8):1–18
21. Simón-Soro A, Mira A (2015) Solving the etiology of dental caries. Trends Microbiol 23(2):76–82
22. Nath SG, Raveendran R (2013) Microbial dysbiosis in periodontitis. J Indian Soc of Periodontol 17(4):543–545
23. Hu X, Zhang Q, Hua H et al (2016) Changes in the salivary microbiota of oral leukoplakia and oral cancer. Oral Oncol 56:e6–e8
24. Ly M, Abeles SR, Boehm TK (2014) Altered oral viral ecology in association with periodontal disease. MBio 5(3):1–13
25. Colson P, Emergents V, Emergence UMR et al (2013) "Megavirales", a proposed new order for eukaryotic nucleocytoplasmic large DNA viruses. Arch Virol 158(12):2517–2521
26. Hoyles L, McCartney AL, Neve H et al (2014) Characterization of virus-like particles associated with the human faecal and caecal microbiota. Res Microbiol 165(10):803–812
27. Kleiner M, Hooper LV, Duerkop BA (2015) Evaluation of methods to purify virus-like particles for metagenomic sequencing of intestinal viromes. BMC Genomics 16:7
28. Castro-Mejía JL, Muhammed MK, Kot W et al (2015) Optimizing protocols for extraction of bacteriophages prior to metagenomic analyses of phage communities in the human gut. Microbiome 3(1):64
29. Thurber RV, Haynes M, Breitbart M et al (2009) Laboratory procedures to generate viral metagenomes. Nat Protoc 4(4): 470–483
30. Steward GF, Culley AI, Mueller JA et al (2013) Are we missing half of the viruses in the ocean? ISME J 7(3):672–679

31. Kim KH, Bae JW (2011) Amplification methods bias metagenomic libraries of uncultured single-stranded and double-stranded DNA viruses. Appl Environ Microbiol 77(21):7663–7668
32. Arriola E, Lambros MBK, Jones C et al (2007) Evaluation of Phi29-based whole-genome amplification for microarray-based comparative genomic hybridisation. Lab Investig 87(1):75–83
33. Yilmaz S, Allgaier M, Hugenholtz P (2010) Multiple displacement amplification compromises quantitative analysis of metagenomes. Nat Methods 7(12):943–944
34. Bredel M, Bredel C, Juric D et al (2005) Amplification of whole tumor genomes and gene-by-gene mapping of genomic aberrations from limited sources of fresh-frozen and paraffin-embedded DNA. J Mol Diagn 7(2):171–182
35. Rhee M, Light YK, Meagher RJ et al (2016) Digital droplet multiple displacement amplification (ddMDA) for whole genome sequencing of limited DNA samples. PLoS One 11(5):e0153699
36. Protocols CSH (2006) Protocols, Cold Spring Harbor. Cold Spring Harbor Protocols
37. Methé B, Nelson KE, Pop M et al (2012) A framework for human microbiome research. Nature 486(7402):215–221
38. Picher ÁJ, Budeus B, Wafzig O et al (2016) TruePrime is a novel method for whole-genome amplification from single cells based on TthPrimPol. Nat Commun 7:13296
39. Direito SOL, Zaura E, Little M et al (2014) Systematic evaluation of bias in microbial community profiles induced by whole genome amplification. Environ Microbiol 16(3):643–657
40. Victoria JG, Kapoor A, Dupuis K et al (2008) Rapid identification of known and new RNA viruses from animal tissues. PLoS Pathog 4(9):e1000163
41. Djikeng A, Halpin R, Kuzmickas R et al (2008) Viral genome sequencing by random priming methods. BMC Genomics 9:1–9
42. Culley AI, Suttle CA, Steward GF (2010) Characterization of the diversity of marine RNA viruses. Man Aquat Vir Ecol:193–201
43. López-Bueno A, Rastrojo A, Peiró R et al (2015) Ecological connectivity shapes quasispecies structure of RNA viruses in an Antarctic lake. Mol Ecol 24(19):4812–4825
44. Rosseel T, Van Borm S, Vandenbussche F et al (2013) The origin of biased sequence depth in sequence-independent nucleic acid amplification and optimization for efficient massive parallel sequencing. PLoS One 8(9):1–9
45. Quail MA, Smith M, Coupland P et al (2012) A tale of three next generation sequencing platforms: comparison of ion torrent, Pacific biosciences and Illumina MiSeq sequencers. BMC Genomics 13(1):341
46. Ari AM (2016) Next-generation sequencing: advantages, disadvantages, and future. In: Hakeem KR, Tombuloğuglu H, Tombuloğuglu G (eds) Plant Omics: trends and applications. Springer International Publishing, New York., Champp, pp 109–135
47. Solonenko S, Ignacio-Espinoza JC, Alberti A et al (2013) Sequencing platform and library preparation choices impact viral metagenomes. BMC Genomics 14:320
48. Bartram AK, Poon C, Neufeld JD (2009) Nucleic acid contamination of glycogen used in nucleic acid precipitation and assessment of linear polyacryl amide as an alternative co-precipitant. BioTechniques 47(6):1019–1022
49. Willner D, Furlan M, Haynes M et al (2009) Metagenomic analysis of respiratory tract DNA viral communities in cystic fibrosis and non-cystic fibrosis individuals. PLoS One 4(10):1–12
50. Klingeborn B, Dinter Z (1972) Equine abortion (herpes) virus: strain differences in susceptibility to inactivation by dithiothreitol. Appl Microbiol 23(6):1121–1124
51. Lopez-Bueno A, Tamames J, Velazquez D et al (2009) High diversity of the viral community from an Antarctic Lake. Science 326(5954):858–861
52. Wei Lim Y, Haynes M, Furlan M et al (2014) Purifying the impure: sequencing Metagenomes and Metatranscriptomes from complex animal-associated samples video link. J Vis Exp 94:1–15
53. Simón-Soro Á, Tomás I, Cabrera-Rubio R et al (2013) Microbial geography of the oral cavity. J Dent Res 92(7):616–621
54. Bergallo M, Costa C, Gribaudo G et al (2006) Evaluation of six methods for extraction and purification of viral DNA from urine and serum samples. New Microbiol 29(2):111–119
55. Lombardi VC, Ruscetti FW, Das Gupta J et al (2009) Detection of an infectious retrovirus, XMRV, in blood cells of patients with chronic fatigue syndrome. Science 326:585–589
56. Naccache SN, Greninger AL, Lee D et al (2013) The perils of pathogen discovery: origin of a novel parvovirus-like hybrid genome traced to nucleic acid extraction spin columns. J Virol 87(22):11966–11977
57. Breitbart M, Salamon P, Andresen B et al (2002) Genomic analysis of uncultured marine viral communities. Proc Natl Acad Sci U S A 99(22):14250–14255

58. Roux S, Solonenko NE, Dang VT et al (2016) Towards quantitative viromics for both double-stranded and single-stranded DNA viruses. PeerJ 4:e2777

59. Duhaime MB, Deng L, Poulos BT et al (2012) Towards quantitative metagenomics of wild viruses and other ultra-low concentration DNA samples: a rigorous assessment and optimization of the linker amplification method. Environ Microbiol 14(9):2526–2537

60. Wu L, Wen C, Qin Y et al (2015) Phasing amplicon sequencing on Illumina Miseq for robust environmental microbial community analysis. BMC Microbiol 15(125):1–12

Part V

Blood–Plasma

Chapter 12

A Method for Isolation of the Virome from Plasma Samples

Pilar Domingo-Calap and Juan Vicente Bou

Abstract

Virome studies are of special interest nowadays. Understanding viral communities in different body compartments will help guide future personalized treatments and to discern between homeostasis and disease. High-throughput sequencing technologies allow us to detect all the nucleic acids present in a sample, including viral ones, by random sequencing. One of the major challenges in virome studies is the correct isolation of the viral nucleic acids from a specific sample. This can be done during the extraction steps (e.g., enrichment of viral capsids), or during the bioinformatic analysis (e.g., removing all human and bacterial sequences). Furthermore, it is an important remark that the treatment of the sample will strongly influence the results. Samples will be treated differently if the ultimate goal is the study of all replicating and encapsidated viruses, including both RNA and DNA ones, if we are only focused on DNA ones, or if we want to analyze all the possible viral nucleic acids in the specific sample, even if the genome is degraded. Here, we present a technique that allows for isolation of viral nucleic acids from plasma samples.

Key words Virus, Plasma, Isolation, Nucleic acids

1 Introduction

Viruses, as the most abundant entities in the biosphere, can be found in almost all environments, including extreme conditions [1]. In humans, viruses can be found in different body compartments, including blood. Identifying viruses present in blood is of special interest, although one of the challenges is to be able to recover all the viruses in a specific sample. Different methodologies have been described to extract viral nucleic acids from blood, plasma, and serum. Some approaches consist in extracting all the nucleic acids directly from the sample, where the proportion of human and bacterial DNA will be high, and then enriching for viral nucleic acids by bioinformatic tools [2]. Others consist in enriching the sample of viruses by ultracentrifugation, which will mainly identify encapsidated viruses [3].

A point to remark in any virome study is that it will be necessary to sequence a large quantity of nucleic acids to be able to detect viruses present at low proportions. Understanding what we

Andrés Moya and Vicente Pérez Brocal (eds.), *The Human Virome: Methods and Protocols*, Methods in Molecular Biology, vol. 1838, https://doi.org/10.1007/978-1-4939-8682-8_12,

want or need to answer is of special interest to determine the methodology to apply. It will be different if we just want to detect viruses (qualitative approach) or if we want to do quantification. For the latter, specific steps should be done (*see* below). For each methodology, a specific threshold of detection should be determined including the specific controls.

Different approaches can be achieved in order to isolate viral nucleic acids from plasma samples. Here, we describe a method which allows for isolation of encapsidated viral RNA and DNA from human plasma samples.

2 Materials

Prepare all the buffers and solutions using ultrapure water, by purifying deionized water (sensitivity of 18.2 MΩ-cm at 25 °C).

Prepare and store all reagents at room temperature, 4 °C or – 20 °C according to indications or manufacturing guidelines. Remember to bring to the correct temperature all reagents prior to use.

Remember to follow waste disposal regulations when disposing waste materials according to your institution's regulation.

2.1 Special Equipment

1. Centrifuge at 12,000 × *g*, 4 °C.
2. Ultracentrifuge at 230,000 × *g*, 4 °C (*see* **Note 1**).
   Remember to precool the centrifuges to 4 °C prior to the experimental steps.
3. Thermocycler.
4. Nanodrop or Qubit 3.0 Fluorometer.

2.2 Recovery of Viral Particles

1. 0.25 N pH 9.5 glycine buffer (*see* **Note 2**).
2. 2× phosphate buffer saline (PBS).
3. 1× PBS.
   Keep glycine buffer and PBS at 4 °C.

2.3 Nuclease Treatment and DNA/RNA Extraction

1. 250 U/μL Benzonase.
2. 100 U/μL Micrococcal nuclease.
3. 20× digestion buffer: 1 M Tris, 100 mM $CaCl_2$, 30 mM pH 8 $MgCl_2$ (*see* **Note 3**).
4. 10 nM ethylenediaminetetraacetic acid (EDTA).
5. High Pure Viral Nucleic Acids Kit (*see* **Note 4**).
   Keep all the reagents at –20 °C, except EDTA at 4 °C.

2.4 Random Amplification

1. Diethylpyrocarbonate (DEPC) treated water.
2. 10 mM dNTP mix.
3. 100 μM random hexamer primer (*see* **Note 5**).

4. 40 U/μL RNaseOUT™ RNase Inhibitor (*see* **Note 6**).
5. 200 U/μL SuperScript IV Reverse Transcriptase (*see* **Note 7**).
6. 5× SuperScript IV Buffer.
7. 0.1 M dithiothreitol (DTT).
8. TruePrime WGA kit (Sygnis, *see* **Note 8**).
   Keep all the reagents at −20 °C.

3 Methods

It is recommended to perform all steps on ice to preserve RNA (*see* **Note 9**).

3.1 Recovery of Viral Particles

1. Centrifuge 20–30 mL of plasma, 140,000 × *g* for 2 h 30 min at 4 °C to pellet the viral particles (*see* **Note 10**).
2. Remove the supernatant with serological pipette from the side opposite to the pellet, avoiding touching the pellet with the viral particles (sometimes pellet cannot be seen). It is recommended to mark the tube in order to know where the pellet should be.
3. Add 5 mL of 0.25 N pH 9.5 glycine buffer, resuspend briefly and incubate for 30 min on ice.
4. Neutralize the reaction with 5 mL of 2× PBS and mix by pipetting up and down.
5. Centrifuge the whole reaction 12,000 × *g* for 15 min at 4 °C to pellet large particles already separated by the glycine buffer.
6. Recover the supernatant and centrifuge 230,000 × *g* for 1 h at 4 °C to pellet the viral particles (*see* **Note 11**).
7. Remove the supernatant with serological pipette from the side opposite to the pellet, avoiding touching the pellet with the viral particles (sometimes pellet cannot be seen) (*see* **Note 12**).
8. Resuspend the pellet in 500 μL of 1× PBS (*see* **Note 13**).

3.2 Nuclease Treatment and DNA/RNA Extraction

1. Mix 187 μL of the previous suspension with 10 μL of 20× digestion buffer, 2 μL of 250 U/μL Benzonase, and 1 μL of 100 U/μL Micrococcal nuclease (*see* **Note 14**).
2. Incubate the mix at 37 °C for 2 h.
3. Put the mix on ice and stop the reaction by adding 10 μL of 10 nM EDTA in order to stop the nuclease activity. Vortex and spin it (*see* **Note 15**).
4. After digestion of free nucleic acid, extract viral nucleic acids with a kit for DNA/RNA purification. High Pure Viral Nucleic Acids Kit is recommended. Use the full volume of the diges-

tion reaction and follow the manufacturing protocol. Elution volume of 25 μL is recommended (*see* **Notes 16–18**).

3.3 Random Amplification

1. Random amplification using random hexamers and SuperScript IV Reverse Transcriptase (*see* **Note 19**). Mix together 0.5 μL of 100 μM random hexamer primer, 1 μL of 10 nM dNTP mix, and 11 μL of sample (vortex and centrifuge briefly). Incubate the mix at 65 °C for 5 min and then 1 min on ice. Independently, mix together 4 μL of 5× SSIV Buffer, 1 μL of 0.1 M DTT, 1 μL of 40 U/μL RNaseOUT, and 1 μL of 200 U/μL SuperScript IV enzyme (vortex and centrifuge briefly). Combine both mixes and incubate at 23 °C for 10 min. Afterward, incubate the reaction at 50 °C for 10 min and 10 min at 80 °C to inactivate the reaction.
2. Mix the RT amplified reaction (20 μL) with 10 μL of non-amplified sample (*see* **Note 20**).
3. Amplify DNA with Phi29 DNA polymerase following the recommendation of the TruePrime WGA kit. Briefly, mix 2.5 μL of sample (**step 2**) and 2.5 μL of buffer D (vortex and spin it). Incubate for 3 min at room temperature. Neutralize the reaction adding 2.5 μL of buffer N. Independently, mix together 26.8 μL of DEPC water, 5 μL of reaction buffer, 5 μL of dNTPs, 5 μL of enzyme 1 (*Tth* PrimPol), and 0.7 μL of enzyme 2 (DNA polymerase Phi29). Combine both mixes and incubate at 30 °C for 3 h, then 10 min at 65 °C to inactivate the enzymes.

3.4 Nucleic Acid Purification and Quantification

1. Purify nucleic acids with a DNA purification kit (*see* **Note 21**). Follow the manufacturer's protocol.
2. Quantify nucleic acid with NanoDrop or Qubit (*see* **Notes 22–26**).

4 Notes

1. These conditions recover the vast majority of viral particles. Ultracentrifugation allows for pelleting small viruses like enteroviruses, which are amongst the smallest viral particles [4].
2. Glycine buffer can be prepared in the lab or bought from Sigma-Aldrich as 0.6 N pH 9.2, and subsequently diluted to 0.25 N and adjusted to pH 9.5 with NaOH.
3. This nuclease cocktail can digest free nucleic acids (RNA or DNA, double-stranded, single-stranded, linear and circular) without protease activity [5].
4. High Pure Viral Nucleic Acids Kit (Roche) is recommended, but it is possible to use other kits to extract both DNA and

RNA from viral particles. For very low sample concentration, PureLink Viral RNA/DNA Kit (Thermo-Fisher Scientific) can be more efficient using carrier RNA, at the expense of reducing the proportion of viral reads. In this case, remember to remove carrier RNA reads during bioinformatics analysis.

5. Thermo-Fisher Scientific provides a 20× random primer mix. Random primers from any other source can be used, but be sure to add at least 1 pmol of random hexamer primers in the retrotranscription reaction (0.5 μL of 100 μM).
6. RNaseOUT protects RNA from possible digestion of RNase A, B, and C. There are other equivalent alternatives, such as Ribolock (Thermo-Fisher Scientific).
7. SuperScript IV Reverse Trancriptase (Thermo-Fisher Scientific) is one of the most sensitive reverse transcriptase enzymes at present. Moloney-Murine Leukemia Virus reverse transcriptase (M-MLV RT, Thermo-Fisher Scientific) can be a cheaper alternative but is far less sensitive.
8. Conventional PCR requires primers. Some authors use a primer with random and fixed regions during retrotranscription (RNA viruses) and/or the first cycles of PCR (DNA viruses). After that, they use the fixed region of primer as target [6–8]. TruePrime WGA kit (Sygnis) allows for amplifying the sample without adding primers.
9. If the aim is to analyze the human virome, including RNA and DNA viruses, all the steps should be done on ice to preserve the RNA. In contrast, to study only DNA viruses, all the steps can be done at room temperature (except where indicated otherwise).
10. For epidemiological assays, plasma volume may come from a pool of several samples (e.g., 2 mL of 10 samples). If it is necessary to obtain the virome of a particular sample, adjust the volume to 20 mL with 1× PBS if necessary.
11. Ultracentrifugation can be modified following Stokes' law. 50,000 × *g* for 1 h is similar to 100,000 × *g* for 30 min.
12. In the step of viral particle recovery, if time is a limiting factor, **steps 3–7** of the recovery of viral particles can be omitted without significantly altering the result.
13. Resuspension volume can be modified. However, reduced volumes could affect the recovery of pelleted viral particles.
14. Nuclease treatment allows for digesting free nucleic acids and increasing the proportion of viral reads (from encapsidated viruses).
15. In order to increase sensitivity of the nuclease treatment, it is possible to increase the reaction volume. Remember to scale

the volume of the digestion buffer and EDTA to maintain the ratio of buffer to EDTA.

16. In the extraction of viral nucleic acids protocol, the use of a carrier RNA is indicated. We recommend not including the carrier RNA in order to increase the proportion of viral reads.
17. Nucleic acid extraction columns can be contaminated with other sequences [9, 10]. Pretreatment of columns with UV light can reduce contamination.
18. Note that after the extraction process, viral nucleic acids are free. Therefore, certain precautionary measures will have to be adopted to protect the RNA from degradation by RNAses: use filter tips, work under UV pretreated laminar flow hood, maintain samples on ice (unless the process requires otherwise), change gloves often, etc.
19. Random hexamer primers are all 6-oligonucleotides combinations that allow for retrotranscribing all RNA nonspecifically.
20. We observe that the detection limit DNA viruses can be decreased due to the retrotranscription reaction (data not shown).
21. DNA Clean & Concentrator −5 (Zymo Research) is recommended. Elution volume should be adapted to the NGS library protocol. If high concentration is desired, this kit allows for eluting with as low as 6 μL.
22. NanoDrop (Thermo-Fisher Scientific) sensitivity is reduced at concentrations below 5 ng/μL. For more sensitive quantification, Qubit 3.0 Fluorometer (Thermo-Fisher Scientific) is recommended (detection limit of 0.1 ng/μL).
23. Viral nucleic acids are already to proceed with the NGS library protocol.
24. Internal controls can help in future analysis. Introducing known viruses that should not be present in the sample at known concentrations can test sensitivity and detect possible biases according to genome type and/or viral particle type (by introducing viruses with different features at the same concentrations). It is also recommended to add a negative control with 1× PBS, in which it is possible to introduce the control viruses as internal controls to check for contamination.
25. Remember that the extraction process reduces the number of nucleic acids in each step, as with any extraction process. Viruses with very low initial proportion can be lost during the extraction if they are under the detection threshold.
26. As an alternative to the proposed method to isolate encapsidated nucleic acids, sometimes it is desirable to detect all viral nucleic acids. In this case, it is recommended to use directly the QIAmp Circulating Nucleic Acid Handbook kit (Qiagen)

from plasma samples. Remember that human and bacterial nucleic acids will comprise the vast majority of the reads, and should be removed in the bioinformatic analysis.

References

1. Suttle CA (2007) Marine viruses--major players in the global ecosystem. Nat Rev Microbiol 5:801–812
2. De Vlaminck I, Khush KK, Strehl C et al (2013) Temporal response of the human virome to immunosuppression and antiviral therapy. Cell 155:1178–1187
3. Kohl C, Brinkmann A, Dabrowski PW et al (2015) Protocol for metagenomic virus detection in clinical specimens. Emerg Infect Dis 21:48–57
4. Pina S, Puig M, Lucena F et al (1998) Viral pollution in the environment and in shellfish: human adenovirus detection by PCR as an index of human viruses. Appl Environ Microbiol 64:3376–3382
5. Conceição-Neto N, Zeller M, Lefrère H et al (2015) Modular aproach to customise sample preparation procedures for viral metagenomics: a reproducible protocol for virome analysis. Sci Rep 5:16532
6. Zou N, Ditty S, Li B et al (2003) Random priming PCR strategy to amplify and clone trace amounts of DNA. BioTechniques 35:503–508
7. Victoria JG, Kapoor A, Dupuis K et al (2008) Rapid identification of known and new RNA viruses from animal tissues. PLoS Pathog 4:e1000163
8. Hang J, Forshey BM, Kochel TJ et al (2012) Random amplification and pyrosequencing for identification of novel viral genome sequences. J Biomol Tech 23:4–10
9. Evans GE, Murdoch DR, Anderson TP et al (2003) Contamination of Qiagen DNA extraction kits with legionella DNA. J Clin Virol 41:3452–3453
10. Erlwein O, Robinson MJ, Dustan S et al (2011) DNA extraction columns contaminated with murine sequences. PLoS One 6:e23484

Chapter 13

Viral Concentration and Amplification from Human Serum Samples Prior to Application of Next-Generation Sequencing Analysis

Xavier Fernandez-Cassi, Marta Rusiñol, and Sandra Martínez-Puchol

Abstract

The protocol presented here allows the isolation, purification, nucleic acid extraction, and amplification of DNA/RNA from viruses present in human sera samples. The method allows the random amplification of the viral genomes present by using a Sequence-Independent, Single-Primer Amplification (SISPA) approach enabling the study of both DNA/RNA viruses. An amplification step is needed, as the concentration of viral DNA/RNA in serum samples is low for direct library preparation. The application of the described protocol guarantees enough randomly amplified double-strand DNA for further library preparation using Nextera XT kit from Illumina.

Key words DNA viruses, RNA viruses, Metagenomics, Blood virome, Sequence-Independent, Single-Primer Amplification (SISPA), Metagenomics

1 Introduction

The application of high-throughput sequencing (HTS) techniques in the virology field has provided us with the detection of known and unknown viruses in a wide range of samples, allowing the study of viral populations circulating through the bloodstream [1, 2]. A general overview of the described methodology is summarized in Fig. 1. The protocol presented in this chapter is based on an ultracentrifugation procedure that allows the concentration of viral particles from previously filtered serum samples. Free viral DNA and non-viral DNA is removed using a DNase treatment. Nucleic acids (NA) of viral origin are extracted from the viral capsids, tagged and amplified by using the Sequence-Independent, Single-Primer Amplification procedure, hereafter known as SISPA. Unlike other molecular techniques, the SISPA method is not hampered by the previous knowledge of the desired targeted virus sequence, allowing the amplification of all viral DNA/RNA

Andrés Moya and Vicente Pérez Brocal (eds.), *The Human Virome: Methods and Protocols*, Methods in Molecular Biology, vol. 1838, https://doi.org/10.1007/978-1-4939-8682-8_13, © Springer Science+Business Media, LLC, part of Springer Nature 2018

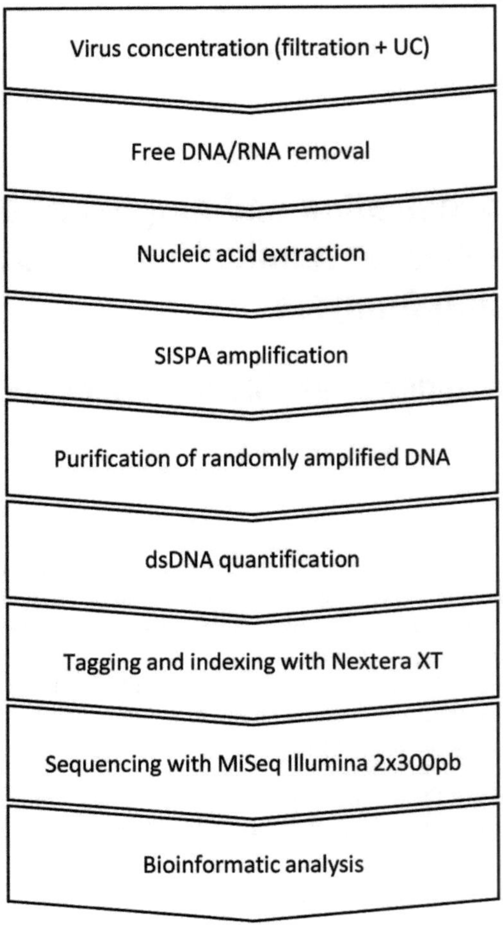

Fig. 1 General overview of the whole procedure

present in a given sample. The SISPA method, developed by Reyes and Kim to randomly amplify DNA [3], has been modified since its original publication to allow the amplification of RNA genomes by using a retrotranscription step [4]. To this day, several random primers have been published in the literature, including 6 or 9 random nucleotides at the 3′-end of the primer [5, 6]. The protocol presented here has used the random nonamer primer A as described by Wang and collaborators [5]. The random tagging and amplification of viral genomes is based on three successive enzymatic reactions: (a) transformation of RNA into cDNA while tagging the RNA genome with the primer A, (b) cDNA complementary strand synthesis and tagging of viral DNA also with primer A, (c) amplification of the tagged viral products by using the constant region of primer A (*see* Fig. 2). As blood samples contain small concentrations of viral DNA/RNA, an amplification step is needed to obtain enough DNA for library preparation. The amplification of dsDNA

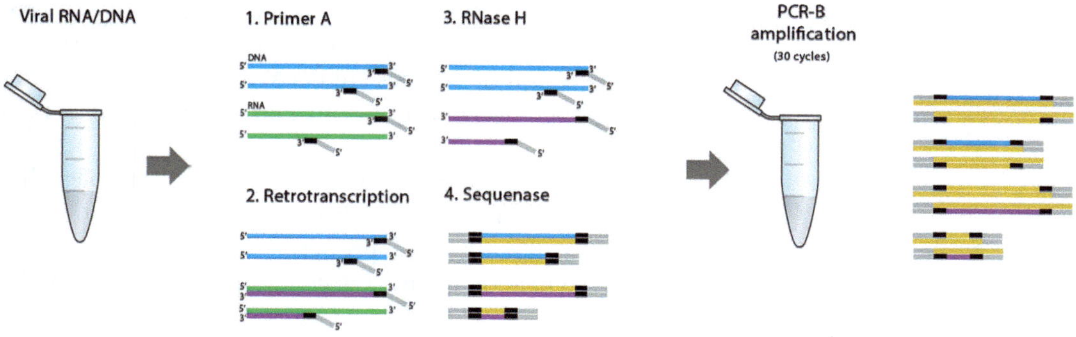

Fig. 2 Overview of the viral DNA/RNA random amplification. DNA in blue, RNA in green, cDNA in purple, double strains generated by sequenase enzyme in yellow, primer-B sequence in gray and primer-A sequence in gray with random nonamer region in black

is done by using a primer B which is complementary to the primer A tagged viral sequences. The number of amplification cycles will also affect the diversity of sequences found in the library. Hence, a minimal number of PCR cycle amplifications should be performed, with 30 cycles being enough to reach a minimum dsDNA input of 1 ng to do the random fragmentation and tagmentation within the library preparation procedure of the Nextera XT kit from Illumina [7]. The protocol for the preparation of viral libraries using Nextera XT kits falls outside the scope of this protocol.

This ultracentrifugation and random amplification protocol has been used to explore the viruses present in serum samples of patients with acute hepatitis with unknown aetiology [1]. The proposed methodology constitutes an excellent tool for diagnostics and monitoring viral evolution, allowing the detection of mixed infections in a single analysis. A similar methodological approach has been followed to study other viromes, including serum [8], faeces [9], and sewage [10], showing the versatility of the method and its usefulness in detecting mixed viral infections, new etiological agents and discovering new viral species [11, 12].

2 Materials

This protocol can be completed within 3 days (*see* **Note 1**). Be careful to avoid the entrance of exogen DNA which might constitute a source of contamination (*see* **Note 2**). Materials used in each step are specified in each section. Unless indicated otherwise, all reagents should be used at room temperature. Plastic materials such as micropipettes, filter tips, and tubes are assumed by the user. All plastic material should be RNase/DNase free. PCR cooler and ice are needed throughout the process to ensure the preservation of DNA/RNA and to guarantee proper performance of the enzymes used throughout the protocol. A benchtop centrifuge able to achieve 20,000 × *g* is required in several steps of the process. Also, a thermoblock system to heat reagents and a vortex to ensure homogenization of samples are needed. Other reagents can be used by adapting the concentrations and conditions provided by the manufacturer.

This protocol is designed to work with serum samples that might contain human or animal pathogens. Please strictly follow biosafety security measures and properly dispose of hazardous waste. Several enzymes or master mixes are used throughout the protocol.

2.1 Cellular Debris and Bacteria Removal by Filtering

1. 0.45 μm diameter low-binding protein filters (*see* **Note 3**).
2. 5 mL sterile plastic tubes.
3. 1.5 mL sterile plastic tubes.
4. Phosphate-Buffered Saline 1× (PBS 1×) at pH 7.4 (*see* **Note 4**).

2.2 Viral Particle Concentration by Ultracentrifugation

For viral concentration, an ultracentrifuge device able to achieve 110,000 × *g* with a compatible rotor will be needed. Tubes able to support 110,000 × *g* speed are also needed.

As an example, in the present protocol a Beckman Coulter ultracentrifuge L-100K with 70Ti rotor has been used. To withstand the applied speed, ultracentrifuge polycarbonate thick wall tubes have been used by filling them with a maximum sample volume of 7 mL.

2.3 DNase Treatment for Free-DNA Removal

1. Ambion TurboDNAse® enzyme (2 U/μL).
2. Ambion TurboDNAse® inactivation agent.
3. Ambion TurboDNASe® buffer 10× .

2.4 Viral Nucleic Acid Extraction

This protocol uses QIAmp® Viral RNA Mini Kit by QIAGEN. Other extraction kits specifically designed for viral DNA/RNA extractions can be used following the instructions provided by the manufacturer.

1. Lysis buffer (AVL) without carrier addition (*see* **Note 5**).
2. Washing buffer AW1.
3. Washing buffer AW2.
4. Ethanol (96–100%). Do not use denatured alcohol, which contains other substances such as methanol or methylethylketone.

2.5 Retrotranscription to cDNA and Tagging of Viral RNA Genomes

This protocol uses a list of commercially available enzymes.

1. SuperScript™ III Reverse Transcriptase (200 U/μL).
2. RNAseOUT™ (40 U/μL).

3. Random primer A (5′-GTTTCCCAGTCACGATCNNN NNNNNNN-3′) at 50 pmol/μL.
4. UltraPure™ DEPC-Treated Water.
5. dNTPs mix: containing a concentration of 10 mM for each of the four nucleotides (A, C, G, and T).
6. RNAse H (5 U/μL).
7. Dithiothreitol (DTT) (100 mM).

2.6 Second Strand Synthesis by Sequenase

This protocol uses a sequenase enzyme, a genetically T7 DNA polymerase used to construct second strand cDNA and to tag DNA genomes. Other alternatives, such as *klenow* DNA polymerase I, can be used (*see* **Note 6**). A programable thermocycle able to perform a ramp of temperature is needed (*see* **Note** 7).

1. Sequenase 2.0 enzyme (13 U/μL).
2. Sequenase buffer.
3. Sequenase dilution buffer.
4. UltraPure™ DEPC-Treated Water.

2.7 PCR-B Amplification of Viral Randomly Tagged dsDNA Chains

This protocol uses AmpliTaq® Gold to do the PCR amplification of viral genomes.

1. UltraPure™ DEPC-Treated Water.
2. Primer B (5′-GTTTCCCAGTCACGATC-3′) at 100 pmols/ μL.
3. AmpliTaq® Gold.
4. $MgCl_2$ (25 mM).
5. 10x PCR Buffer.

2.8 Agarose Electrophoresis Gel

1. Agarose Molecular Biology reagent.
2. Tris-Borate-EDTA 10× (TBE 10×). To prepare TBE 10×: Weigh 121.1 g Tris base, 61.8 g Boric acid, and 7.4 g EDTA and pour them into 1000 mL of deionized water.
3. Tris-Borate-EDTA 1× (TBE 1×). To prepare TBE 1×: pour 100 mL of TBE 10× into 900 mL of deionized water.
4. GelRed® Nucleic acid Stain.
5. Blue/Orange Loading Dye 6×.
6. GeneRuler 100 bp Plus DNA Ladder, ready-to-use.

2.9 Cleaning Viral Randomly Amplified PCR Products and Concentration of Viral dsDNA

Alternative kits with lower elution volumes (15–20 μL) can be used following the manufacturer's instructions (*see* **Note 8**).

1. DNA Clean & Concentrator.
2. UltraPure™ DEPC-Treated Water.

2.10 Fluorometric Quantification

A minimal dsDNA concentration of ≥1 ng/μLof randomly amplified viral genomes is needed for library preparation. To accurately quantify the dsDNA produced during PCR amplification a fluorometric quantification system must be used. In the present protocol, a Qubit® fluorometer and the Qubit® dsDNA HS Assay Kit from Life technologies have been used (*see* **Note 9**).

Real-time PCR (polypropylene) 0.5 mL tubes compatible with fluorometric measurement are needed.

2.11 Library Preparation by Using Illumina Nextera XT Kits

The amount of dsDNA obtained at the end of this protocol should be enough for library preparation using Nextera XT kit. The minimal amount of dsDNA needed for Nextera XT kit for library preparation is ≥1 ng. Nextera XT library preparation instructions provided by the manufacturer should be followed (https://support.illumina.com/sequencing/sequencing_kits/nextera_xt_dna_kit/documentation.html).

3 Methods

3.1 Sample Pooling and Cellular Debris and Bacteria Removal by Filtering

In the present protocol, serum samples from individual patients are combined to obtain different pooled samples according to clinical criteria. Pooled samples are obtained by adding 500 μL for each sample. In order to remove cellular debris and possible microbial contaminants, pooled sera samples are passed through 0.45 μm pore size low-binding protein filters.

1. Pass 1 mL of sterile PBS 1× through the filter in order to condition before serum filtering (*see* **Note 3**).
2. Prepare a collector plastic tube and pass the 5 mL of pooled sera sample through the filter and collect the filtrate.
3. Keep the filtrate at 4 °C until further use. Unless samples are going to be processed on the same day, keep them at −80 °C.

3.2 Viral Particle Concentration by Ultracentrifugation

Viral particles present in the pooled sera samples are pelleted by ultracentrifugation and suspended into 500 μL of PBS.

1. Place each filtered pooled sera sample into an ultracentrifuge tube.
2. In case of processing an odd number of samples, balance the rotor by placing a tube with the same volume of water or PBS 1×. Check that both tubes contain exactly the same weight before starting the ultracentrifugation process.
3. Ultracentrifuge for 1 h at 100,000 ×*g* and 4 °C (*see* **Note 10**).
4. Discard the supernatant and suspend the pellets, which contain the viral particles, by using 500 μL of PBS 1×. The pellet is not always visible, be sure to suspend it carefully.
5. Store viral concentrates at −80 °C until further use.

3.3 DNase Treatment for Free-DNA Removal

Viral nucleic acids, including DNA and RNA, are isolated from 300 μL of the obtained viral concentrate. Before viral capsid disruption, non-viral free DNA from the host is removed by the application of a DNase treatment. The remaining 200 μL can be used afterward for molecular testing to confirm the presence of viral DNA detected in serum samples by conventional molecular methods.

Add 50 μL of the viral concentrate to six different 0.5 mL sterile plastic tubes (*see* **Note 11**).

1. Add to each of the 6 tubes the following reagents and vortex vigorously:
   - 10 μL of TurboDNase® (2 U/μL).
   - 10 μL 10× TurboDNase® reaction Buffer.
2. Incubate for 1 h –37 °C by using a Thermoblock to ensure homogeneous heating of the sample. After the first 30 min of incubation add 3 μL of turbo DNase enzyme per tube. Incubate for another half an hour (*see* **Note 12**). Regular agitation during incubation is recommended.
3. Collect in a single 1.5 mL sterile plastic tube the DNase-treated viral concentrate from the 6 tubes.
4. Inactivate the DNase treatment by adding 44 μL of the inactivation reagent (*see* **Note 13**). Once the reagent is added, the tube has to be inverted a couple of times in order to mix the reagent with the tube content. Do not vortex the sample once inactivation agent is added. Incubate the mixture for 5 min at room temperature (22–26 °C).
5. Centrifuge the tubes at 10,000 × *g* for 1.5 min. After centrifugation a white pellet is observed. Keep the supernatant containing the viral particles without disturbing the pellet and place it in a new collection tube. Discard the visible pellet.

3.4 Viral Nucleic Acid Extraction

A nucleic acid extraction kit is used to extract viral DNA/RNA. In this protocol QIAmp® Viral RNA Mini Kit from QIAGEN is used. To extract the nucleic acids 280 μL of the viral DNase-treated concentrate are used. Remember to include a negative control for the whole process.

1. Pipet 1120 μL of AVL Buffer into a 5.0 mL centrifuge tube.
2. Add 280 μL of viral concentrate to the 1120 μL of Buffer AVL. Mix the tube by vortexing for 15 s.
3. Incubate at room temperature (15–25 °C) for 10 min to disrupt the viral nucleocapsids (*see* **Note 14**).
4. Briefly, spin the tubes to remove drops from inside of the lid.
5. Add 1120 μL of ethanol (96%–100%) to the sample.

6. Mix by vortexing for 15 s. Spin the tubes briefly to remove drops from the inside of the lid.
7. Carefully apply 630 μL of the solution from the **step 5** to a QIAamp® Mini spin column (in a 2 mL collection tube) without wetting the rim. Centrifuge at 6000 × *g* for 1 min. Place the QIAamp spin column into a clean 2 mL collection tube and discard the tube containing the filtrate.
8. Repeat the **step** 7. Be sure all sample volume has gone through the filter.
9. Add 500 μL of washing buffer AW1. Centrifuge at 6000 × *g* for 1 min. Keep the spin column, and discard the collection tube and the filtrate.
10. Add 500 μL of washing buffer AW2. Centrifuge at 20,000 × *g* for 3 min. Keep the spin column and discard the collection tube and the filtrate.
11. Put a new collection tube and centrifuge at 20,000 × *g* for 1 min. Keep the spin column.
12. Place the QIAamp® Mini Spin Column into a clean 1.5 mL centrifuge tube (collection tube). Open the column and add 60 μL of AVE. Incubate the Spin Column for 2 min and centrifuge at 6000 × *g* for 1 min.
13. Store the eluted NA at −80 °C for further molecular analysis.

3.5 Retrotranscription to cDNA and Tagging of Viral RNA Genomes

The cDNA step is prepared in two tubes: the master mix tube and the sample tube. The master mix tube will contain all the common components for the retrotranscription reaction in order to minimize pipetting errors. The sample mix, unique for any sera sample, will contain primers, dNTPs and the specific DNA/RNA extracted from the sample.

1. For each sample prepare a Sample tube with 1 μL of dNTPs Mix, 6 μL of nucleic acids extraction (*see* **Note 15**), 1 μL of random primer A. Adjust the water volume to a maximum final volume of 13 μL (i.e., If using 6 μL of nucleic acids extraction, add 5 μL of UltraPure™ DEPC-Treated Water).
2. Denature the RNA and primer by incubating for 5 min at 65 °C.
3. Chill on ice for 5 min to break secondary structures of RNA.
4. Prepare a master mix tube. Make the calculations for *n* + 2 samples in order to have enough volume. Include a negative control for the whole process. Add the following reagents per sample: 2 μL of 5× RT buffer SuperScript™ III, 4 μL of 25 mM $MgCl_2$, 2 μL of 0.1 M DTT, 1 μL of RNase™ Out, and 1 μL of SuperScript™ III enzyme.
5. Homogenize the master mix tube by vortexing.

6. Add 10 μL of the master mix to each sample tube.
7. Place the sample tubes into the PCR thermocycler and run the RT program as follows: 25 °C for 10 min, 50 °C for 50 min, 85 °C for 5 min.
8. Place the PCR tubes in a cooler.
9. Add 1 μL of RNase H to each sample tube in order to denature RNA-DNA hybrid structures which could affect subsequent reactions. Then, incubate the PCR tubes in a PCR thermocycler for 20 min at 37 °C. Until further use keep the tubes at 4 °C.

3.6 Second Strand Synthesis by Sequenase

Second strand cDNA is constructed in the PCR tubes resulting from the retrotranscription.

1. Incubate the sample tubes at 95 °C for 5 min in a PCR thermocycler.
2. Rapidly chill all the sample tubes in ice or in a cooler for 5 min.
3. Prepare a second master mix tube (master mix 2). Make the calculations for $n + 2$ samples in order to have enough volume. Master mix 2 contains the following reagents for each sample: 2 μL of sequenase buffer, 0.3 μL of sequenase enzyme, and 7.7 μL of UltraPure™ DEPC-Treated Water.
4. Add 10 μL of master mix 2 into each sample tube.
5. Place the PCR tubes in a thermocycler (*see* **Note 7**). The second strand synthesis needs a temperature ramp: Ramp 1 °C every 18 s, 37° for 8 min, 94 °C for 2 min, and 10 °C for 5 min.
6. A second step using sequenase to entirely complete the second strand is performed. In this second reaction, the enzyme is diluted using the enzyme dilution reagent provided. Prepare a master mix 3. Again, make the calculations for n + 2 samples. Master mix 3 reagents are 0.3 μL sequenase enzyme and 0.9 μL sequenase dilution buffer.
7. Add to each tube 1.2 μL of master mix 3.
8. Repeat the program described in **step 5**, increasing the 94 °C time from 2 to 8 min.
9. Keep the primer A tagged viral dsDNA sequences at 4 °C.

3.7 PCR-B Amplification of Viral Randomly Tagged dsDNA Chains

The PCR master mix should be prepared in a separate area to avoid contamination with other PCR amplicons. Template from PCR reaction should be added in a separate room. Keep the reagents and master mix chilled on ice. Amounts provided are per reaction tube. Prepare 2 reaction PCR tubes for each tested sample. A negative PCR reaction tube for this step should be included. Consider an extra sample to account for pipetting error ($n + 1$). Add the following reagents to obtain a 90 μL PCR-B master mix for the sample.

1. PCR-B master mix preparation: add 8 μL of $MgCl_2$, 10 μL of PCR Buffer10×, 1 μL dNTPs, 1 μL of Taq DNA polymerase, 1 μL of Primer, and 69 μL of UltraPure™ DEPC-Treated Water.
2. In a separate room add the Sequenase template. For each sample tested, two parallel PCR reactions are performed by using 20 out of 30 μL of sequenase template. Keep the spare volume for an extra PCR-B reaction, if needed (*see* **Note 16**).
3. Place the tubes in the thermocycler and follow the program: 95 °C-10 min, followed by 30 cycles at 94 °C for 30 s, 40 °C for 30 s, 50 °C for 30 s and 72 °C for 1 min. Finally, an extension at 72 °C for10 min and 4 °C-hold

3.8 Electrophoresis Gel

Run an agarose electrophoresis to visualize the smears produced in the PCR-B amplification.

1. Measure 2 g of agarose.
2. Mix the agarose powder with 100 mL 1 × TBE in a microwavable flask.
3. Microwave for 2–3 min until the agarose is completely dissolved (usually above 83 °C) and let it cool down 5 min to about 50 °C.
4. Add GelRed® safe fluorescent dye for staining dsDNA to a final concentration of 10 μg/100 mL.
5. Pour the agarose into a gel tray with the well comb in place and let it sit at room temperature for 5–10 min until it has completely solidified.
6. Place the agarose gel into the gel box (electrophoresis unit).
7. Add a molecular weight ladder into the first and the final lanes of the gel.
8. Mix 10 μL of the PCR-B products with 3 μL of the loading buffer and carefully load the mixture into the additional wells of the gel.
9. Run the gel at 90–110 V until the dye line is approximately 75–80% of the way down the gel.
10. Turn OFF power, disconnect the electrodes from the power source, remove the gel from the gel box, and visualize your DNA fragments using any standard UV transilluminator (*see* **Note 17**).

3.9 Cleaning Viral Randomly Amplified PCR Products and Concentration of Viral dsDNA

Excess of primers and dNTPs not used during the PCR amplification should be removed to obtain dsDNA clean PCR-B products. The two PCR-B tubes obtained per each sample are cleaned and concentrated in a single reaction tube.

1. Pool the content of the two PCR-B tubes of each sample into a 1.5 mL plastic tube. A final volume of 190 μL is expected.
2. Add 1 mL of DNA binding buffer within the sample tube (a DNA binding proportion 5:1 with the sample is expected).
3. Briefly vortex the mixture and transfer the mix into a Zymo column.
4. Centrifuge at 14,000 × *g* for 30 s.
5. Discard the eluate and place the column in a new collector tube.
6. Add 200 μL of DNA wash buffer into the column.
7. Centrifuge at 14,000 × *g* for 30 s.
8. Discard the eluate and place the column in a new collector tube.
9. Repeat **steps 5** and **6**.
10. Discard the eluate and place the column in a new collector tube.
11. Preheat to 70 °C one plastic tube containing DNase/RNase free water (*see* **Note 18**).
12. Add 15 μL pre-warmed at 70 °C water DNase/RNase free to the column and place a new plastic tube under the column.
13. Incubate for 5 min.
14. Centrifuge at 14,000 × *g* for 1 min.
15. Keep the eluate that contains the purified viral dsDNA for library preparation.

3.10 Fluorometric Quantification

Use a Qubit® Fluorometer to ensure that a final concentration of 1 ng of viral dsDNA has been obtained (*see* **Note 19**).

1. Prepare two assay tubes for the standard curve. Prepare as many assay tubes as there are samples to be tested.
2. Prepare 200 μL of qubit working solution buffer for each standard and sample to be tested. The Qubit® working solution consists of 1 μL of the qubit reagent and 199 μL Qubit Buffer per reaction. Ensure the mixture of both components by vortexing 10 s.

3. For the two standard tubes place 190 μL of Qubit® working solution per tube and add 10 μL of the standard solution 1 or 2. To quantify the DNA concentration from samples, place 199 μL of Qubit® working solution and 1 μL of purified PCR-B product.
4. Vortex the tubes for 2–3 s and wait a couple of min at room temperature before taking the reading.
5. Insert the tubes in the Qubit® Fluorometer and turn on the device.
6. On the Home screen, select the assay for *dsDNA measure*. Select the *high sensitivity assay*.
7. Select *read the standards*. Place standard 1 tube into the sample chamber and select *read standard*. Do the same with standard 2 tube. After observing the calibration curve the Qubit® device is ready to read the samples.
8. On the *read standard* screen, select *run samples*. On the sample volume screen, select the sample volume as *1 μL*. Select the units expressed as *ng/μL*.
9. Insert a sample tube into the sample chamber, close the lid, and then select *read tube* to get the concentration of your sample.

Samples will be ready for library preparation with Nextera XT kit.

4 Notes

1. The authors suggest completing the protocol within 3 consecutive days. On day 1 viral particles can be filtered through 0.45 μm, concentrated by ultracentrifugation and DNase treated. On day 2, nucleic acids can be extracted, RNA retrotranscribed and DNA/RNA tagged and amplified with random primers A and B, respectively. On day 3, dsDNA amplified DNA will be purified, dsDNA concentration checked, and quality of DNA assessed by gel electrophoresis.
2. All steps use Microbial DNA-Free Water and DNA-free reagents. All plastic tubes should be DNA-Free and UV-treated to destroy traces of DNA present due to the sensitivity of the methodology to possible DNA contaminants. Try to use, as far as possible, new processing material when following this protocol and disposable plastic material in order to avoid, as far as possible, the introduction of external contaminating DNA.
3. Prior to filtering sera samples, add 1 mL of deionized DNA-Free water to each filter unit. This step will moisten the filter and substantially minimize sample loss during the filtration procedure.

4. If possible, try to use certified commercially available PBS 1× buffer that is RNase-free and tested for contaminating nonspecific endonuclease, exonuclease, and RNase activity. Avoid the use of autoclaved material, as well as autoclaved water, as it can also constitute a source of contaminating DNA. DNA has been proved to resist the autoclaving procedure [13].
5. When using silica technology columns for viral NA extraction, the addition of carrier enhances the recovery of NA from samples. Carrier can prevent that small amounts of target NA present in the sample can bound irretrievably to the materials used during the column based extraction. The carrier is an external source of nucleic acids (>200 nt) that binds into the silica membrane. In order not to interfere with sequencing, no carrier is added.
6. *Klenow* polymerase is an alternative to sequenase to construct the second strand and tag DNA genomes. If sequenase is not available, use *klenow* polymerase as follows: add 0.6 μL of dNTPs mix (10 mM) and 0.5 μL *klenow* fragment exo per sample tube. Incubate the tubes at 37 °C for 60 min. Finally, inactivate *klenow* by incubating at 75 °C for 10 min.
7. If a thermocycler with the capacity to perform a temperature ramp is not available for sequenase reaction, a conventional thermocycler may be used by raising the temperature from 10 °C to 37 °C, increasing by 1 °C every 18 s.
8. PCR products need to be purified of excess dNTPs and primers not used during the PCR amplification step. The low elution volume used will help in concentrating amplified PCR products in order to have enough concentration (1 ng/μL) for library preparation.
9. The Qubit® dsDNA HS Assay Kit is designed to provide accurate quantifications of dsDNA when initial sample concentrations range from 10 pg/μL to 100 ng/μL. Alternatives to Qubit® Fluorometer and the Qubit® dsDNA HS Assay Kit such as PicoGreen® DNA Quantitation Reagent for DNA can be also used to assess dsDNA concentrations.
10. Precool to 4 °C the ultracentrifuge and rotors before use to guarantee a 4 °C temperature during all the ultracentrifugation process.
11. This step is performed in six different tubes to ensure that the DNase enzyme has a better interaction with free DNA present in the processed sample and ensure its optimal degradation. After this treatment, any contaminant DNA cannot be eliminated. The protein capsid of viral particles protects them from the nuclease activity of DNase enzyme, preserving viral DNA and RNA from degradation.

12. To ensure complete degradation of free DNA and to rule possible inhibition or under performance of the DNase used, 3 μL of Turbo DNase are added to boost the reaction.
13. The DNase inactivation buffer has a high density. Be sure that the inactivation reaction buffer is vortexed prior to its addition to each DNase-treated sample. This will avoid the addition of insufficient reagent and will ensure the end of the DNase reaction.
14. Viral particle lysis is complete after 10 min incubation at room temperature. Viral particles are lysed under denaturing conditions provided by Buffer AVL. Longer incubation times have no effect on the yield or quality of the purified RNA. This step also inactivates RNases to ensure intact viral RNA isolation. Potentially infectious agents present in the samples lose their infectivity after nucleocapsid disruption.
15. This protocol is designed to work efficiently with 1–500 ng of RNA/DNA input material. The amount of viral DNA/RNA added to the reaction can be upscaled to 11 μL of nucleic acid extraction by reducing the water volume added in the same proportion, helping to achieve the desired concentrations of viral DNA/RNA for library preparation. A second option to achieve viral DNA/RNA concentrations is to prepare 2 different retrotranscription reaction tubes per sample by using 6 μL as described in the protocol. This second option will also guarantee the availability of a second cDNA tube in case of failure of the retrotranscription enzyme.
16. Only 20 μL of sequenase reaction will be used to perform 2 PCR-B reactions. The remaining volume can be used for an extra PCR-B reaction in case two amplification PCR-B are not enough to achieve the minimum concentration for library preparation. To achieve enough DNA for library preparation, it is preferable to perform three PCR-B amplification reactions rather than increasing the number of PCR cycles applied during the amplification. This PCR step might introduce bias by amplifying the most abundant genomes and producing %GC bias [14]. Less abundant genomes might not be sequenced or may be underrepresented [15]. A balance between the number of cycle amplifications and DNA concentrations for library preparation must be achieved. The application of a higher number of PCR cycles can produce higher DNA concentrations but decreases the estimated viral richness of the samples [16].
17. A DNA smear must be visualized in the SISPA amplified samples. It is preferable for the smear to present a higher DNA concentration between 600 and 1000 base pairs. A negative sample followed through all the process (from extraction to PCR-B amplification) must be loaded into the agarose gel to

evaluate the level of contamination. The negative control smear must be invisible or clearly lighter than the smear of serum samples. Molecular biology reagents are produced and purified from bacteria or their infecting viruses (bacteriophages or phages). Therefore, the inclusion of contaminant DNA during the production of these enzymes cannot be avoided and can explain the smear in negative controls. If the smear is clearly visible in the negative control it can suggest a DNA contamination due to improper sample treatment. If dsDNA concentration is achieved for library preparation, it is strongly recommended to run Mi-seq negative controls from your samples to assess DNA contaminants during the procedure.

18. The elution of DNA from the column depends on pH and temperature conditions. Make sure the water's pH is >6.0. The total DNA yield may be improved by eluting the DNA from the column with preheated water at 70 °C and increasing the elution time to 5 min. By doing so, larger amplicons (6–10 kbs) will be eluted with higher efficiency ensuring the inclusion of larger randomly amplified viral fragments to our libraries.

19. Store the Qubit dye and the Qubit buffer at room temperature. dsDNA standards have to be stored at 4 °C. Ensure that all reagents are at room temperature when preparing the assay. To ensure that reagents are at room temperature and the dsDNA measure obtained by the Qubit is accurate, it is recommended to remove the standards from the fridge 20–30 min before preparing the standard reactions.

References

1. Gonzales-Gustavson E, Timoneda N, Fernandez-Cassi X et al (2017) Identification of sapovirus GV.2, astrovirus VA3 and novel anelloviruses in serum from patients with acute hepatitis of unknown aetiology. PLoS One 12:e0185911
2. Moustafa A, Xie C, Kirkness E et al (2017) The blood DNA virome in 8,000 humans. PLoS Pathog 13:e1006292
3. Reyes GR, Kim JP (1991) Sequence-independent, single-primer amplification (SISPA) of complex DNA populations. Mol Cell Probes 5:473–481
4. Froussard P (1993) rPCR: a powerful tool for random amplification of whole RNA sequences. Genome Res 2:185–190
5. Wang D, Urisman A, Liu YT et al (2003) Viral discovery and sequence recovery using DNA microarrays. PLoS Biol 1:E2
6. Allander T, Emerson SU, Engle RE, Purcell RH, Bukh J (2001) A virus discovery method incorporating DNase treatment and its application to the identification of two bovine parvovirus species. Proc Natl Acad Sci 98:11609–11614
7. Illumina (2014) Nextera® XT DNA library preparation kit. Reporter:8–11
8. Wylie TN, Wylie KM, Herter BN et al (2015) Enhanced virome sequencing using targeted sequence capture. Genome Res 25:1910–1920
9. Kapoor A, Li L, Victoria J et al (2009) Multiple novel astrovirus species in human stool. J Gen Virol 90:2965–2972
10. Fernandez-Cassi X, Timoneda N, Martínez-Puchol S et al (2017) Metagenomics for the study of viruses in urban sewage as a tool for public health surveillance. Sci Total Environ 618:870–880
11. Thorburn F, Bennett S, Modha S et al (2015) The use of next generation sequencing in the

diagnosis and typing of respiratory infections. J Clin Virol 69:96–100

12. Blomström AL, Widén F, Hammer AS et al (2010) Detection of a novel astrovirus in brain tissue of mink suffering from shaking mink syndrome by use of viral metagenomics. J Clin Microbiol 48:4392–4396
13. Choi WS, Rodríguez RA, Sobsey MD (2014) Persistence of viral genomes after autoclaving. J Virol Methods 198:37–40
14. Duhaime MB, Deng L, Poulos BT et al (2012) Towards quantitative metagenomics of wild viruses and other ultra-low concentration DNA samples: a rigorous assessment and optimization of the linker amplification method. Environ Microbiol 14:2526–2537
15. Karlsson OE, Belák S, Granberg F (2013) The effect of preprocessing by sequence-independent, single-primer amplification (SISPA) on metagenomic detection of viruses. Biosecur Bioterror 11(Suppl 1):S227–S234
16. Fernandez-Cassi X, Timoneda N, Gonzales-Gustavson E et al (2017) A metagenomic assessment of viral contamination on fresh parsley plants irrigated with fecally tainted river water. Int J Food Microbiol 257:80–90

Part VI

Urinary Tract

Chapter 14

Identification and Quantification of DNA Viral Populations in Human Urine Using Next-Generation Sequencing Approaches

Tasha M. Santiago-Rodriguez

Abstract

The human urinary virome is comprised by diverse communities of both eukaryotic viruses and bacteriophages. Identification and quantification of human urinary viruses was limited to PCR-based and culture methods. Detection of viruses in human urine has long been associated with disease, but advances in next-generation sequencing have shown that it harbors robust viral communities. Human urinary viral communities are mainly comprised by bacteriophages and Human Papillomaviruses (HPVs) that remain to be characterized in association with health and diverse urinary disease states. In the present chapter, the methods to identify and quantify DNA viral populations in human urine using next-generation sequencing are described.

Key words Bacteriophages, Human papillomaviruses, Microbiome, Urinary diseases, Virome

1 Introduction

Human urine has long been assumed to be sterile until reaching the urethra, and that the detection of microorganisms is a sign of disease. The presence of microbial communities in human urine may have been overlooked in the past due to the application of culture techniques, many of which may have only captured viable and culturable bacteria including *Escherichia coli* and *Enterococcus* spp.; yet, an increasing number of studies applying next-generation sequencing approaches have supported the presence of bacterial communities in human urine in association with health [1], and also several different disease states including urinary tract infections (UTIs) [2–4], overactive bladder [5], urgency urinary incontinence [6, 7], interstitial cystitis [8], and type-2 diabetes [9]. Bacterial communities in human urine also prompted the investigation of viral communities and how they may be implicated in

Andrés Moya and Vicente Pérez Brocal (eds.), *The Human Virome: Methods and Protocols*, Methods in Molecular Biology, vol. 1838, https://doi.org/10.1007/978-1-4939-8682-8_14,

maintaining health or potentially promoting diverse urinary disease states.

Viral communities in human urine are usually associated with disease because most studies have focused on eukaryotic viruses, which only represent a small proportion of the human virome [4]. Several studies have highlighted the importance of detecting eukaryotic viruses in human urine including Human Papillomaviruses (HPVs) [10], BK polyomavirus, JC virus and Torque teno virus [11] to potentially circumvent more invasive procedures for their detection [10]; however, bacteriophages comprise the majority of viruses in human urine [4]. Bacteriophages in human urine may be mostly lysogenic, infecting bacteria from the Proteobacteria, Firmicutes, Bacteroidetes, and Verrucomicrobia phyla. A fraction of a healthy human urinary virome is also comprised by HPVs that are not implicated with cervical and rectal cancers [4].

In the present chapter, the methods for the detection of viruses in human urine are detailed as described in [4]. While the methods by Santiago-Rodriguez and collaborators, 2015 are detailed, researchers are encouraged to test different methods to identify and quantify DNA viral communities in human urine. Generally, the main workflow to identify and quantify viruses from human urine should be to: (1) remove human and bacterial cells, (2) isolate and concentrate viral particles, (3) remove free nucleic acids, (4) extract viral DNA, (5) amplify viral DNA, (6) prepare viral DNA for whole-genome sequencing, and (7) analyze results. Notably, with modifications including, but not limited to RNA isolation, construction of cDNA libraries, RNA viral communities can potentially be studied in human urine.

2 Materials

2.1 Pre-processing Urine Samples for Isolation of Virions

1. SM buffer: Dissolve 5.8 g of NaCl and 2.0 g of $MgSO_4•6H_2O$ in 800 mL of H_2O. Add 50 mL of 1 M Tris–HCl (pH 7.5). Adjust the volume to 1 L with H_2O. Sterilize the buffer by autoclaving for 20 min. Store at room temperature.
2. 0.45 μm filters.
3. 0.2 μm filters.
4. Syringes.
5. Sterile tubes.

2.2 Purification of Virions Using Caesium Chloride (CsCl) Gradient Ultracentrifugation

1. CsCl solutions: Prepare three CsCl solutions consisting of 1.70 g/mL, 1.50 g/mL, and 1.35 g/mL. For this, dissolve CsCl in sterile SM buffer. Filter-sterilize solutions.
2. Ultracentrifuge tubes.

3. Serological pipets.
4. Ultracentrifuge.
5. Sterile syringes.
6. Sterile gauge needles.
7. Sterile tubes.

2.3 Concentration of Virions Using Protein Filters

1. Amicon UltraCentrifugal filters.
2. SM buffer.
3. Microcentrifuge tubes.
4. Microcentrifuge.

2.4 DNA Extraction from Recovered Virions

1. SM buffer.
2. DNase I.
3. Microcentrifuge.
4. Microcentrifuge tubes.
5. Waterbath or heatblock.
6. QIAamp UltraSens Virus Kit (*see* **Note 1**).
7. NanoDrop spectrophotometer.

2.5 Amplification of Viral DNA

1. Illustra GenomiPhi V2 DNA Amplification Kit.
2. Microcentrifuge tubes.
3. Waterbath or heatblock.

3 Methods

3.1 Pre-processing Urine Samples for Isolation of Virions

1. Collect urine samples aseptically and process within 24 h after collection (*see* **Note 2**).
2. Attach a 0.45 μm filter to a sterile syringe and pipette 500 μL of sterile SM buffer to the syringe, and pass it through the filter. This will increase the amount of sample flow through.
3. Pipette urine samples through the 0.45 μm filter to remove cellular debris and collect in sterile tube (*see* **Note 3**).
4. Attach a 0.2 μm filter to a sterile syringe and pipette 500 μL of sterile SM buffer to the syringe, and pass it through the filter. This will increase the amount of sample flow through.
5. Filter filtrate from above through a 0.2 μm filter and collect in a sterile tube.

3.2 Purification of Virions Using Caesium Chloride (CsCl) Gradient Ultracentrifugation

Although a number of methods have been tested for the isolation of viruses from diverse sample types [12], CsCl gradient ultracentrifugation has been used for the isolation of virions from human urine [4].

1. Add 0.5 g/mL CsCl to the sequentially filtered urine sample.
2. Invert the tube gently until completely dissolved.
3. Prepare and label clean ultracentrifuge tubes for each sequentially filtered urine sample.
4. Carefully, pipette the CsCl layers into sterile and labeled ultracentrifuge tubes. Add the solution of 1.70 g/mL to the bottom of the ultracentrifuge tube; followed by the 1.50 g/mL, and the 1.35 g/mL solutions, respectively (*see* **Note 4**).
5. Carefully pipette the sequentially filtered urine sample containing 0.5 g/mL CsCl on the top of the CsCl gradient.
6. Carefully place the tubes in the ultracentrifuge rotor and place it in ultracentrifuge.
7. Centrifuge at 33,550 × *g* at 4 ° C for 2 h.
8. Carefully remove the rotor and tubes to avoid disturbing the gradient.
9. Puncture the ultracentrifuge tube with the undisturbed CsCl gradient using the appropriate sterile gauge needle, and remove the layer corresponding to the most commonly known bacteriophages (*see* Fig. 1).
10. Place bacteriophage layer in a sterile and labeled tube.

3.3 Concentration of Virions Using Protein Filters

1. Add recovered bacteriophage fraction from each sample to an Amicon Ultra Centrifugal Filter for protein purification and concentration. Use one filter per sample.
2. Centrifuge Amicon Ultra Centrifugal Filter at 3,100 × *g* for 5 min.
3. Discard flow through and repeat until loading the remaining sample.
4. Place filter in a clean collection tube.
5. Add SM buffer to the Amicon Ultra Centrifugal Filter (*see* **Note 5**).
6. Flick the tube 4 to 5 times to disperse the buffer across the filter.
7. Invert Amicon Ultra Centrifugal Filter and centrifuge at 3,100 × g for 5 min.
8. Immediately proceed to DNA extraction or store flow through at −20 °C for not more than 48 h.

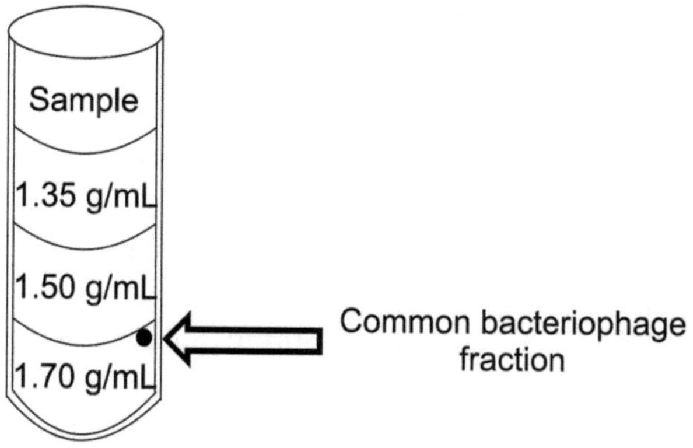

Fig. 1 Cesium Chloride ultracentrifuge tube. Puncture with the undisturbed CsCl gradient using the appropriate sterile gauge needle to remove the layer corresponding to the most commonly known bacteriophages

3.4 DNA Extraction from Recovered Virions

A number of commercially available DNA extraction kits have been designed for DNA viruses. There are also a number of non-commercial options that have been widely applied for DNA extraction of virions including phenol-chloroform followed by ethanol precipitation, and cetyl trimethyl ammonium bromide or CTAB. Here we described the method used by Santiago-Rodriguez and collaborators 2015 using the QIAamp® UltraSens Virus Kit with modifications described below.

1. Add 950 μL of SM buffer to the virions recovered from **step 3** in Subheading 3.
2. Add 10 U 10^{-1} DNase I to the collected purified virions and incubate at 37 °C for 10 min.
3. Transfer the virions (1 mL) into a 2 mL microcentrifuge tube and pipet 800 μL of buffer AC to the sample.
4. Pipet 5.6 μL carrier RNA solution into the tube lid.
5. Invert the microcentrifuge tube 3 times and vortex microcentrifuge tube for 10 s.
6. Incubate at room temperature for 10 min.
7. Centrifuge at 600 × *g* for 3 min.
8. Carefully remove and discard the supernatant (*see* **Note 6**).
9. Pipet 300 μL of Buffer AR warmed to 60 °C (*see* **Note 7**).
10. Pipet 20 μL of proteinase K and vortex until the pellet is completely resuspended.
11. Incubate at 40 °C for 10 min and quick spin the microcentrifuge tube.
12. Pipet 300 μL of Buffer AB and vortex to mix.
13. Quick spin the microcentrifuge tube and transfer the 700 μL lysate to a spin column.

14. Centrifuge at 3,640 × *g* for 1 min.
15. Discard flow through and place spin column into a new 2 mL collection tube.
16. Pipet 500 μL Buffer AW1 and centrifuge at 6,219 × *g* for 1 min.
17. Discard flow through and place spin column into a new 2 mL collection tube.
18. Pipet 500 μL Buffer AW2 and centrifuge at 16,872 × g for 3 min.
19. Discard flow through and place the spin column into a 1.5 mL microcentrifuge tube.
20. Pipet 30 μL of Buffer AVE to the membrane of the spin column and incubate at room temperature for 3–5 min.
21. Centrifuge at 4,842 × *g* for 1 min.
22. Repeat by adding 30 μL of Buffer AVE to the membrane of the spin column and incubate at room temperature for 3–5 min.
23. Centrifuge at 4,842 × *g* for 1 min.
24. Analyze DNA quantity and quality using a NanoDrop spectrophotometer.

3.5 Amplification of Viral DNA

Whole genome sequencing of virions from diverse sample types may be preceded by amplification of nucleic acids using commercially available kits. The method used by Santiago-Rodriguez and collaborators, 2015 is described below. The method of viral DNA amplification was performed using illustra GenomiPhi V2 DNA Amplification Kit as follows:

1. Dilute viral DNA from **step 4** in Subheading 3 to 10 ng/μL.
2. Add 9 μL of Sample Buffer to 1 μL of 10 ng template DNA.
3. Heat the samples to 95 °C for 3 min, then cool to 4 °C on ice.
4. Combine 9 μL of Reaction Buffer with 1 μL of Enzyme Mix for each sample on ice.
5. Add 10 μL of prepared master mix from above to each cooled sample on ice.
6. Incubate the samples at 30 °C overnight.
7. Heat the samples to 65 °C for 10 min.
8. Cool samples to 4 °C.
9. Analyze DNA quantity and quality using a NanoDrop spectrophotometer.
10. Store at −20 °C.

3.6 Library Preparation and Sequencing

Researchers are encouraged to utilize the library preparation and sequencing method of their choice. The Ion Torrent PGM was utilized in the paper by Santiago-Rodriguez and collaborators 2015.

3.7 Data Analysis

Although the methods for data analysis of urinary viromes are presented as those described by Santiago-Rodriguez and collaborators 2015, there are a number of bioinformatic tools specifically developed to characterize viral metagenomes including Viral Informatics Resource for Metagenome Exploration (VIROME) [13], Metavir [14], and Metavir 2 [15] that could also be potentially used to characterize urinary viromes.

3.7.1 Pre-processing and Quality Filtering Sequence Files

1. Check quality of fastq files (*see* **Note 8**).
2. Convert files from.fastq format to.fasta format (*see* **Note 9**).
3. Remove low complexity reads, homopolymers, and those with great length variation (*see* **Note 10**).
4. Run a BLASTN analysis of the reads against the Ribosomal Database Project (RDP) 16S rRNA database available at http://rdp.cme.msu.edu/misc/resources.jsp, and the human reference database available at ftp://ftp.ncbi.nlm.nih.gov/genomes/H_sapiens/ to screen for contaminating bacterial and human nucleic acids (*see* **Note 11**).
5. Remove any reads homologous to bacterial and/or human sequences (e-value $<10^{-5}$) from **step 4** prior to further analysis (*see* **Note 12**).

3.7.2 Reads Assembly and Annotation

1. Assemble pre-processed reads from **step 1** in **step 7** in Subheading 3, based on 98% identity, and with a minimum of 50% read overlap (*see* **Note 13**).
2. Run a BLASTX analysis with the assembled reads against the NCBI NR database (ftp://ftp.ncbi.nlm.nih.gov/blast/db/) with an e-value cutoff value of 10^{-5}.
3. Parse BLASTX results to determine viral taxonomy (*see* **Note 14**).
4. Parse BLASTX results to determine functional categories for known viral genes including replication, structural, transposition, restriction/modification, hypothetical, and other genes previously found in viruses for which the e-value was at least 10^{-5} (*see* **Note 14**).
5. Compile data in a spreadsheet to determine the relative proportions of assembled contigs that contained viral homologues to particular taxonomy and/or functional categories for known viral genes.

3.7.3 Mapping of Reads to Virus Reference Genomes

Mapping virome reads to reference virus genomes can provide insights into the genome architecture, gene location, and potential function of viruses, and can also aid in the reconstruction of potentially novel viruses isolated from human urine. As with many of the methods described above, a number of bioinformatic tools have been developed to map sequences against reference genomes including Bowtie. In the paper by Santiago-Rodriguez and collaborators 2015, CLC Genomics Workbench was used to map virome reads against reference databases as described below:

1. Download virus databases from www.phantome.org and ftp://ftp.ncbi.nih.gov/genomes/Viruses/.
2. Map virome reads or contigs to both databases using a 98% identity over a minimum of 50% of the read length (*see* **Note 15**).

4 Notes

1. While alternative methods exist for the extraction of viral DNA, these have not been tested for viruses isolated from human urine. Other more affordable methods including phenol:chloroform, followed by ethanol precipitation, and CTAB have been widely tested for the isolation of viral nucleic acids from diverse sample types [12].
2. Samples can be processed up to 72 h after collection without altering results. Store samples at 4 °C. DO NOT freeze samples.
3. Sample quantity will depend on availability. Approximately 3 mL of sample was processed in the study by [4].
4. If using 4 mL ultracentrifuge tubes, 1 mL of each CsCl solution should be used, and 1 mL of the sample.
5. Quantity will depend on the amount of starting material. 50 μL have been used for 3 mL of urine sample.
6. While the pellet is expected at the bottom of the microcentrifuge tube, a portion of the pellet can attach to the sides of the microcentrifuge walls. Be sure to not touch the walls of the microcentrifuge tubes.
7. Adding the warmed AR buffer directly to the microcentrifuge tube walls may result in higher DNA yields.
8. It is recommended to use FastQC or similar tools to check the quality of the sequences. FastQC can be downloaded from http://www.bioinformatics.babraham.ac.uk/projects/fastqc/. There is also an online version available at http://galaxy.dbcls.jp/tool_runner?tool_id=fastqc.

9. CLC Genomics Workbench was used in the paper by [4], but other non-commercial options, such as QIIME, can be used using the script split_libraries_fastq.py (http://qiime.org/scripts/split_libraries_fastq.html). The resulting sequence file will be in the.fna format.
10. Low complexity reads, homopolymers, reads with length variation (<50 nucleotides or >300 nucleotides) and those with ambiguous characters can be removed using Ion (454) Assist. The tool was developed in Dr. David Pride laboratory at the University of California San Diego. The tool has a GUI interface and is freely available to download from http://www.thepridelaboratory.org/software.html. A user manual is available to download http://nebula.wsimg.com/bee1214315ac49ccb39f9c41119157cc?AccessKeyId=FF87537C73100FA9AEBE&disposition=0&alloworigin=1
11. For more information of running a local BLAST visit https://www.ncbi.nlm.nih.gov/guide/howto/run-blast-local/. BLAST analyses can also be done online using tools such as BLASTer in DiaGrid (https://diagrid.org/), and with commercially available tools such as CLC Genomics Workbench.
12. While researchers often develop proprietary scripts, a number of bioinformatic tools have been developed to filter bacterial and human sequences from viromes including ViromeScan http://sourceforge.net/projects/viromescan/ and Ion (454) Assist. Commercially available tools such as CLC Genomics Workbench can be used to map virome reads against 16S rRNA gene and human databases. The program provides a file with sequences not matching the databases. The resulting file represents the virome reads that can be used for further analyses.
13. A number of tools have been used to assemble virome reads including CLC-Genomics Workbench, SPAdes and ABySS. Researchers may prefer certain assemblers over others based on availability and performance, including time to obtain results.
14. While researchers often develop proprietary scripts to parse a BLAST output, a number of tools are available which are python-based, such as BLAST parser (http://kirill-kryukov.com/study/tools/blast-parser/). Other researchers have made publicly available their scripts, such as that included in the Ion (454) Assist tool.
15. In the paper by Santiago-Rodriguez and collaborators 2015, mapping of urinary virome reads against the described databases, resulted in a number of reads mapping to uncharacterized HPVs. The study validated part of the data by designing primers specific for the HPV sequences identified in the viromes.

References

1. Lewis DA, Brown R, Williams J et al (2013) The human urinary microbiome; bacterial DNA in voided urine of asymptomatic adults. Front Cell Infect Microbiol 3:41
2. Fouts DE, Pieper R, Szpakowski S et al (2012) Integrated next-generation sequencing of 16S rDNA and metaproteomics differentiate the healthy urine microbiome from asymptomatic bacteriuria in neuropathic bladder associated with spinal cord injury. J Transl Med 10(1):174
3. Detweiler K, Mayers D, Fletcher SG (2015) Bacteruria and urinary tract infections in the elderly. Urol Clin North Am 42(4):561–568
4. Santiago-Rodriguez TM, Ly M, Bonilla N et al (2015) The human urine virome in association with urinary tract infections. Front Microbiol 6:14
5. Hilt EE, McKinley K, Pearce MM et al (2014) Urine is not sterile: use of enhanced urine culture techniques to detect resident bacterial flora in the adult female bladder. J Clin Microbiol 52(3):871–876
6. Pearce MM, Hilt EE, Rosenfeld AB et al (2014) The female urinary microbiome: a comparison of women with and without urgency urinary incontinence. MBio 5(4):e01283–e01214
7. Pearce MM, Zilliox MJ, Rosenfeld AB et al (2015) The female urinary microbiome in urgency urinary incontinence. Am J Obstet Gynecol 213(3):347.e341–347.e311
8. Siddiqui H, Lagesen K, Nederbragt AJ et al (2012) Alterations of microbiota in urine from women with interstitial cystitis. BMC Microbiol 12(1):205
9. Liu F, Ling Z, Xiao Y et al (2016) Dysbiosis of urinary microbiota is positively correlated with type 2 diabetes mellitus. Oncotarget 8(3):3798–3381
10. D'Hauwers K, Tjalma W (2009) Screening for human papillomavirus: is urine useful? Indian J Cancer 46(3):190
11. Rani A, Ranjan R, McGee HS et al (2016) A diverse virome in kidney transplant patients contains multiple viral subtypes with distinct polymorphisms. Sci Rep 6:33327
12. Kleiner M, Hooper LV, Duerkop BA (2015) Evaluation of methods to purify virus-like particles for metagenomic sequencing of intestinal viromes. BMC Genomics 16:7
13. Wommack KE, Bhavsar J, Polson SW et al (2012) VIROME: a standard operating procedure for analysis of viral metagenome sequences. Stand Genomic Sci 6(3):427–439
14. Roux S, Faubladier M, Mahul A et al (2011) Metavir: a web server dedicated to virome analysis. Bioinformatics 27(21):3074–3075
15. Roux S, Tournayre J, Mahul A et al (2014) Metavir 2: new tools for viral metagenome comparison and assembled virome analysis. BMC Bioinf 15:76

Part VII

Bioinformatics

Chapter 15

Diversity Analysis in Viral Metagenomes

Jorge Francisco Vázquez-Castellanos

Abstract

Viruses are the most abundant and diverse biological entity in the earth. Nowadays, there are several viral metagenomes from different ecological niches which have been used to characterize new viral particles and to determine their diversity. However, viral metagenomic data have the disadvantage to be high-dimensional compositional and sparse. This type of data renders many of the conventional multivariate statistical analyses inoperative. Fortunately, different libraries and statistical packages have been developed to deal with this problem and perform the different ecological and statistical analyses. In the present chapter, it is analyzed simulated viral metagenomes, based on real human gut-associated viral metagenomes, using different R and python packages. The example presented here includes the estimation and comparison of different indexes of diversity, evenness, and richness; perform different ordination and statistical analysis using different dissimilarity metrics; determine the optimal cluster configuration and perform biomarker discovery. The scripts and the simulated datasets are in https://github.com/jorgevazcast/Viromic-diversity

Key words Ordination analysis, Viral metagenomics, Alpha diversity, Beta diversity, Clustering and biomarker discovery

1 Introduction

Biodiversity refers to the wide variety of living beings on Earth and the natural patterns that make it up. The biodiversity describes the diversity from genes to species and studies the number of different populations, species and, the durable interactions between species and their environment [1]. Species diversity refers to the number of species that are found within a landscape. The study of diversity can be divided into three distinct types: alpha, beta. and gamma diversity. The alpha diversity measures how many different species are and how evenly distributed are in a certain habitat, study-area, or sample, it is also known as the within-sample diversity [2]. The alpha diversity can be measured using indexes such as the Shannon index [2, 3] or the effective number of species [4]. Moreover, different approximations such as the species richness, which quantifies the number of different species within a habitat (ex: the Chao1 [5, 6] and the ACE index [6]) and the species evenness, which is the

Andrés Moya and Vicente Pérez Brocal (eds.), *The Human Virome: Methods and Protocols*, Methods in Molecular Biology, vol. 1838, https://doi.org/10.1007/978-1-4939-8682-8_15,

similarity in the relative abundance of the species (ex: Pielou's J′ evenness [7]), are complementary to describe alpha diversity. The beta diversity was defined as "the extent of change in community composition, or degree of community differentiation, in relation to a complex-gradient of the environment, or a pattern of environments" [8]. It is also known as the between-sample diversity and measures the differences between two different habitats or samples. Commonly different dissimilarity distance and index are used to study beta diversity. In metagenomic analysis the most common are the Bray-Curtis dissimilarity (quantifies dissimilarities based on the species counts), the Jaccard distance (quantifies the differences between samples based on the presence/absence species), the UniFrac distance (quantifies dissimilarities incorporating phylogenetic information), the Euclidean distance after a proper data normalization among several others [8, 9]. Finally, the gamma diversity is the total species diversity in a landscape, this encompasses alpha and beta diversity [8].

Diversity analysis is fundamental for the study of the microbiome so that there are different libraries and specialized packages mainly focused on the study of bacterial diversity [10, 11]. This is because bacterial microbiomes have been the most abundant in the last decade. Recent massive metagenomic studies [12, 13] have increased the knowledge of viral diversity and distribution, have discovered a large number of possible new viral species, and have elucidated their putative hosts. In this scenario, the study of the viral metagenomics is a useful tool for studying viral diversity since it allows characterizing phages that infect non-culturable bacterial species. However, there are technical limitations which complicate the study of viral diversity such as the poor taxonomic annotation [14], the lack of phylogenetic gene markers, and the high diversity. The sum of these generates a large number of sequences without correct taxonomic annotations and the identification of viral particles which are specific for only a few samples. These methodological biases generate sparse data [15], which refer to data matrices that contain a large proportion of zeros. Sparse data is problematic for several ordination and regression analyses that assume normality. Moreover, viral metagenomes also suffer from the high-dimensional compositional data problem which is common for all the metagenomic datasets. Compositional data renders many standard multivariate statistical methods inappropriate or inapplicable [15, 16]. Fortunately, there are different statistical packages which can correctly deal with this type of problems and make estimates of the diversity in viral metagenomics as well as the detection of biomarkers related to different conditions. The purpose of this chapter is to show these techniques and provide a practical example.

2 Materials

1. PC with Linux 3 or higher (tested on Ubuntu 16.04 and 14.04 distributions), alternatively, macOS or Windows 10 (with Windows subsystem for Linux enabled).
2. Although all these analyses can be performed on a computer with at least two processors and two gigabytes of RAM, I recommend using a computer that has at least 8 processors (64 bits) and 8 gigabytes of RAM. This will help to greatly reduce the computation times in several of the analyses.
3. R version 3.2.3 or above.
4. python 2.7 or above and below python 3.
5. Inkscape https://inkscape.org/en/.
6. Bash 4 or equivalent shell.
7. Stable internet connection (>1 Mbps is preferable).

3 Methods

All scripts and datasets are accessible at https://github.com/jorgevazcast/Viromic-diversity.

3.1 Install R and the Packages Necessary for the Data Analysis

3.1.1 Open a Bash Terminal and Install R

Important: the "$" symbol indicates the prompt and is not part of the command and you should not copy-paste these instructions into the prompt line.

```
$ sudo add-apt-repository 'deb [arch=amd64,i386]
https://cran.rstudio.com/bin/linux/ubuntu xenial/'
(see Note 1)

$sudo apt-get update
$ sudo apt-get install r-base r-base-dev
```

3.1.2 Open R and Install the Necessary Packages for the Current Analysis

R is open by typing R in the Bash prompt.

```
$ R
```

If the installation was successful, this should appear on the screen.

R version 3.2.3 (2015-12-10) -- "Wooden Christmas-Tree"

Copyright (C) 2015 The R Foundation for Statistical Computing.

Platform: x86_64-pc-linux-gnu (64-bit).

R is free software and comes with ABSOLUTELY NO WARRANTY.

You are welcome to redistribute it under certain conditions.
Type 'license()' or 'licence()' for distribution details.
Natural language support but running in an English locale.
R is a collaborative project with many contributors.
Type 'contributors()' for more information and

'citation()' on how to cite R or R packages in publications.
Type 'demo()' for some demos, 'help()' for on-line help, or
'help.start()' for an HTML browser interface to help.
Type 'q()' to quit R.
>

The ">" symbol indicates that now you are into the R environment (*see* **Note 2**)

3.1.3 Install the Necessary Libraries that Are Going to Be Used in the Analysis

Important: the ">" symbol indicates the R prompt and is not part of the command and you should not copy-paste these instructions into the promt line.

```
>install.packages(c("fossil","vegan","FactoClass","scatterplot3d","MASS","cluster","gplots","RcolorBrewer","pvclust","splines", "stats4", "survival", "mvtnorm", "modeltools", "coin","fields") ) (see Note 3)
>source("https://bioconductor.org/biocLite.R")
>biocLite("metagenomeSeq") (see Note 4)
```

3.1.4 Quit R

```
>q()
```

3.2 Install Python and the Packages Necessary for the Data Analysis

```
$ python -V
```

3.2.1 Check the Python Version

```
Python 3.6.0 :: Continuum Analytics, Inc.
```

3.2.2 If Python 2.7 Is Not Installed, Install Using This Command

```
$sudo apt-get install python2.7
```

3.2.3 Install the Packages

Install the following packages: rpy2, argparse, matplotlib, numpy, pandas, scikit-learn, scipy

The packages can be installed via pip install

```
$ python 2.7 pip install scipy # (see Note 5)
```

or via Anaconda

```
$conda install numpy # (see Note 6)
```

Repeat one of these actions for all the before-mentioned packages.

3.3 Download and Install the Biomarker Discovery Software

3.2.4 Download the LEfSe Software (See Note 7)

Download the LEfSe (Download repository) from this website:

https://bitbucket.org/nsegata/lefse/downloads/

Move the software to the desired directory (etc., software LEfSe). For this example, the directory will be named software.

```
$ mkdir ~/software/
$ unzip ~/Downloads/nsegata-lefse-54694b4b0d9e.zip -d ~/software/ # (see Note 8)
```

3.4 Create the Working Directory and Download the Data

```
$ mkdir example
$ cd example
$ wget https://raw.githubusercontent.com/jorgevaz-cast/Viromic-diversity/master/simulated_viral_data.tsv
$ wget https://raw.githubusercontent.com/jorgevaz-cast/Viromic-diversity/master/viral_diversity.R
$ wget https://raw.githubusercontent.com/jorgevaz-cast/Viromic-diversity/master/supplementary_diversity_functions.R
```

3.5 Read Data and Normalization

3.5.1 Open the R Environment

```
$ R
```

3.5.2 Open All the Libraries That Are Going to Be Used

```
> library(fossil)
> library(vegan)
> library(cluster)
> library(gplots)
> library (RColorBrewer)
> library(pvclust)
> library(metagenomeSeq)
>library(RCurl)
> source("./supplementary_diversity_functions.R")
```

3.5.3 Read the Data Table of the Relative Abundances of the Viral "Species" (see Note 9)

Function to read the file that is in the table format

```
> abundance.table<-read.table(file="simulated_viral_data.tsv", header=T, row.names = 1
,dec=".",sep="\t")
```

Function to visualize the data

```
> head(abundance.table)
```

3.5.4 Create the Metadata Table (see Note 10)

The metadata simulates two conditions, the "C" and the "H" condition

```
>metadata<-cbind(colnames(abundance.table),c(rep("H",100) , rep("C",100) ))
>rownames(metadata)<-colnames(abundance.table)
>metadata<-as.data.frame(metadata)
>colnames(metadata)<-c("samle_name","HC")
> head( metadata )
```

3.5.5 Exclude Rare Species That Could Be Taxonomic Binning Artifacts

The "zero.vec" contains the percentage of "0" for each sample

```
>zero.vec<-apply(abundance.table,1,function(x){
length(x[x==0])/length(x)} )
```

Exclude all those virus like particles whose viral like particles are above the 10% of the samples

```
> abundance.table.nr.sp<-abundance.table[zero.
vec<0.1,] # (see Note 11)
```

Compare the dimension of both matrices

```
>dim(abundance.table)

>dim(abundance.table.nr.sp)
```

3.5.6 Plot the Rarefaction Curves Using the Function "Rarecurve" from the Vegan Library

Estimates the minimum number of sequences (*see* Fig. 1)

```
>min.number.secs<-min(apply(abundance.table.
nr.sp,2,sum))
>#png(filename = " Fig1_rarefaction_curves.png",
width = 600, height = 600) # (see Note 12)
>pdf(file = "Fig1_rarefaction_curves.pdf")
> rarecurve(t(abundance.table.nr.sp), step = 200,
sample = min.number.secs, col = "blue", cex = 0.6)
# (see Note 13)
>dev.off() # (see Note 12)
```

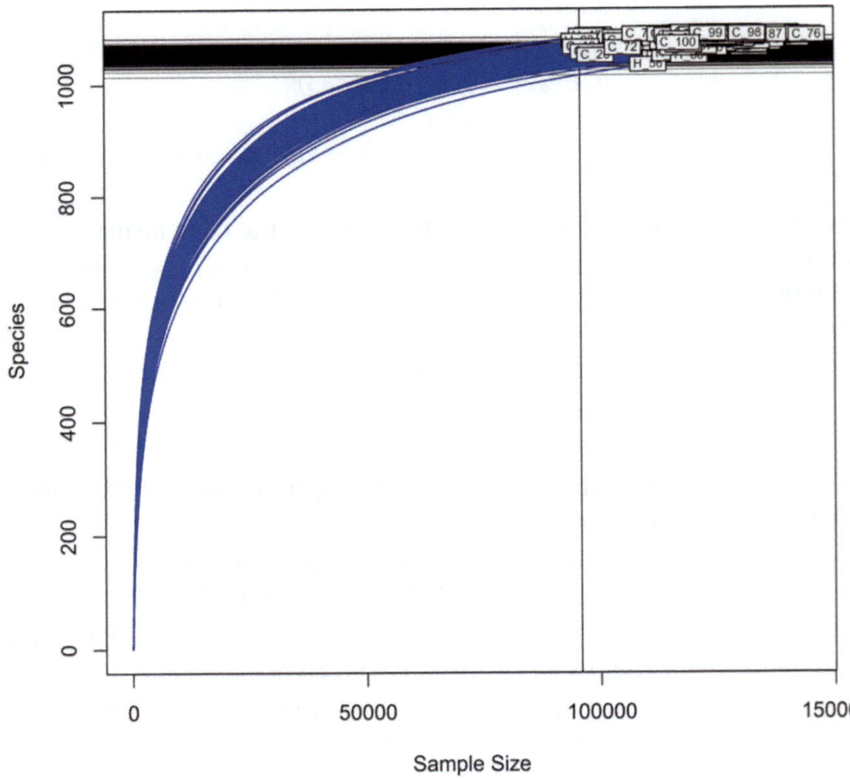

Fig. 1 Rarefaction curves. Rarefactions curves for the 200 samples. All samples reach the diversity plateau which indicates that the sequencing effort has saturated the diversity curves. The vertical line points to the sample with the lesser number of reads. Horizontal lines represent the rarefied species richness

3.5.7 Rarefaction: Performs a Rarefaction to the Sample with the Minimum Number of Reads

Perform the rarefaction

```
>viromic.rar<-rrarefy(t(abundance.table.nr.sp),
min(apply(abundance.table.nr.sp,2,sum))) # (see
Note 14)
```

Verify the output

```
>apply(viromic.rar,1,sum)
```

3.5.8 Standardization: Performs the CSS Standardization

It standardizes each one of the samples by its respective CSS value. The "cumNormMat" function is allocated into the metagenomeSeq library.

```
>viromic.CSS<-t(cumNormMat(as.matrix(abundance.table.
nr.sp))) #(see Note 15)
```

3.6 Alpha Diversity Analyses (See Note 16)

Alpha diversity indexes

```
>Shannon <- diversity(viromic.rar) # Shannon in-
dex, the function is coded in the "vegan" library

>Specnumber <- specnumber(viromic.rar) # Expected
number of species, the function is coded in the
"vegan" library

>Pielou <- pielou(viromic.rar) # Pielou index, the
function is coded in the supplementary functions
file "supplementary_diversity_functions.R"

>Chao1<-apply(viromic.rar,1, chao1) # Chao1 index,
the function is coded in the "fossil" library
>ACE<-apply(viromic.rar,1, ACE) # ACE index, the
function is coded in the "fossil" library
```

3.6.1 Plot the Boxplot of the Different Diversity Index. The Significance Between Condition Was Given by the Wilcoxon Test (See Fig. 2)

Divide the plotting window in three rows and two columns

```
>pdf("Fig2_Alpha_diversity_index_comparison.pdf")

>par(mfrow=c(3,2))
```

Calculate the p-value of the Wilcoxon test

```
>p.val<-round(wilcox.test(Shannon~metadata[,2])$p.
value, digits=3)

> boxplot(Shannon ~ metadata[,2],col=c("red","dark
green"), main="Shannon index",sub=
paste("WT p-val =",p.val), ylab=
"Shannon index") # Represents the data in a box-
plot # (see Note 17)

>legend("bottomleft", legend=c("C", "H"), pch=16,
col=c("red", "darkgreen"), cex=0.7,box.lty=0,
title="Legend") # Add the plot legend

>p.val<-round(wilcox.test(Specnum
ber~metadata[,2])$p.value, digits=3)

> boxplot(Specnumber ~ metadata[,2],col=("
red","darkgreen"), main="Expected number of
species",sub=paste("WT p-val =",p.val), ylab="Num
species")
```

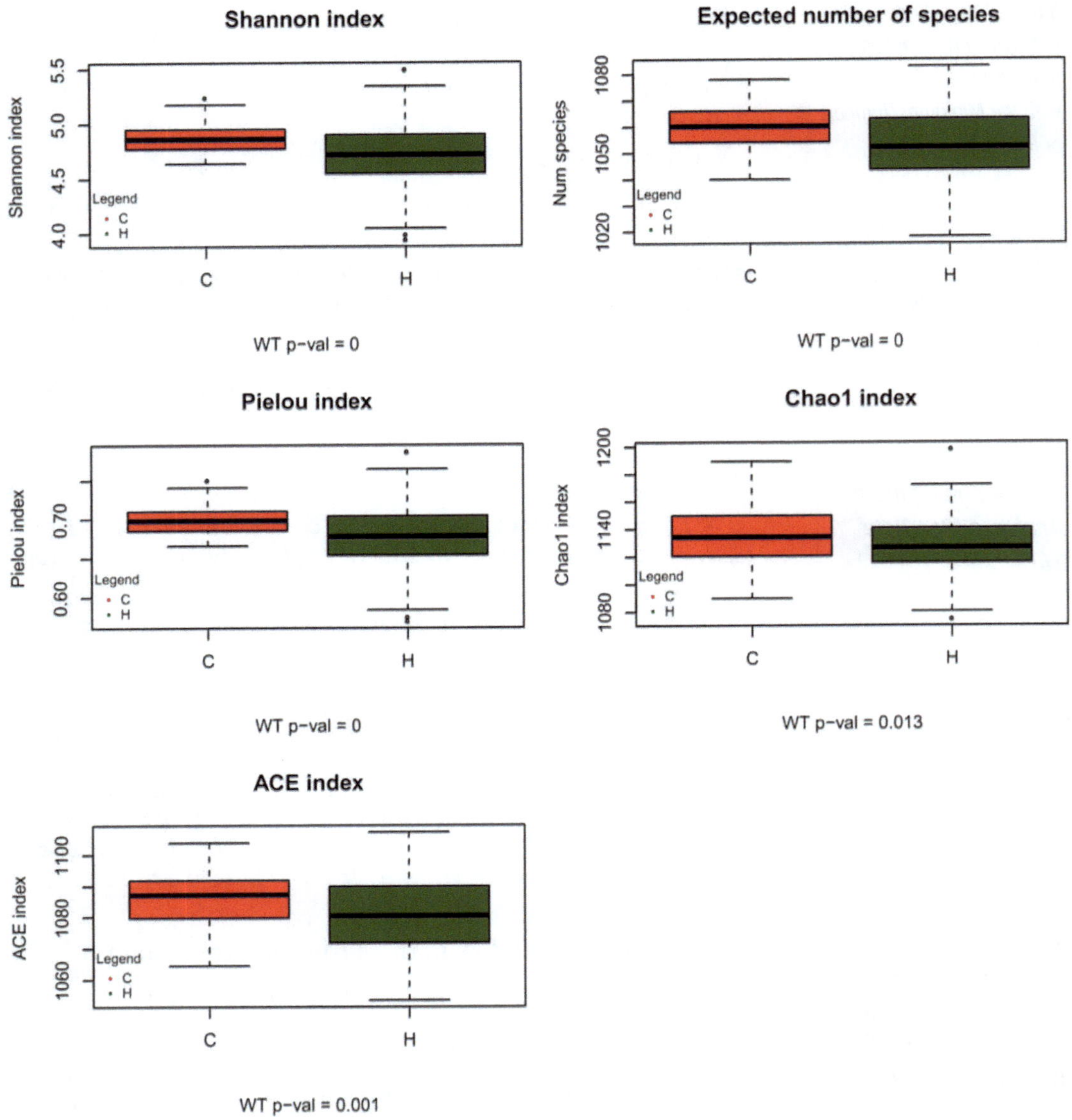

Fig. 2 Alpha diversity index comparison. Boxplot of the five different diversity indexes used in the analysis for both conditions H (green) and C (red). The statistical significance between both conditions is determined by the Wilcoxon test

```
>legend("bottomleft", legend=c("C", "H"), pch=16,
col=c("red", "darkgreen"), cex=0.7,box.lty=0,
title="Legend") # Add the plot legend

>p.val<-round(wilcox.test(Pielou~metadata[,2])$p.
value, digits=3)

> boxplot(Pielou ~ metadata[,2],col=c("red","dark
green"), main="Pielou index",sub=paste("WT p-val
=",p.val), ylab="Pielou index")

>legend("bottomleft", legend=c("C", "H"), pch=16,
col=c("red", "darkgreen"), cex=0.7,box.lty=0,
title="Legend") # Add the plot legend
```

```
>p.val<-round(wilcox.test(Chao1~metadata[,2])$p.value, digits=3)
> boxplot(Chao1 ~ metadata[,2],col=c("red","dark green"), main="Chao1 index",sub=paste("WT p-val =",p.val), ylab="Chao1 index")
>legend("bottomleft", legend=c("C", "H"), pch=16, col=c("red", "darkgreen"), cex=0.7,box.lty=0, title="Legend") # Add the plot legend
>p.val<-round(wilcox.test(ACE~metadata[,2])$p.value, digits=3)
> boxplot(ACE ~ metadata[,2],col=c("red","darkgreen"), main="ACE index",sub=paste("WT p-val =",p.val), ylab="ACE index")
>legend("bottomleft", legend=c("C", "H"), pch=16, col=c("red", "darkgreen"), cex=0.7,box.lty=0, title="Legend") # Add the plot legend
>dev.off()
```

3.7 Beta Diversity Analyses (See Note 18)

3.7.1 Calculate a Principal Coordinate Analysis (PCoA) Using the Bray-Curtis Index

```
>PCoA<-capscale(viromic.CSS ~ 1, metaMDS = TRUE,sqrt.dist= TRUE) # (see Note 19)
>pdf("Fig3_Default_PCoA.pdf")
```

Plot the PCoA using default parameters (*see* Fig. 3)

```
> plot(PCoA)
>dev.off()
```

3.7.2 Other Dissimilarity Index to Calculate the Beta Diversity (see Note 20)

Dissimilarity index to calculate the beta diversity:

(a) Bray-Curtis

```
>rar.bray.dist<-vegdist(viromic.rar)
>CSS.bray.dist<-vegdist(viromic.CSS)
```

(b) Hellinger norm and Euclidean distance

```
>rar.hell.dist<-vegdist(decostand(viromic.rar,method="hell"),method="euclidean")
>CSS.hell.dist<-vegdist(decostand(viromic.CSS,method="hell"),method="euclidean")
```

(c) Jaccard distance

```
>viromic.rar.binary<-viromic.rar
>viromic.rar.binary[viromic.rar.binary>0]=1 # (see Note 21)
>rar.jaccard.dist<-vegdist(viromic.rar.binary, method = "jaccard", binary = TRUE)
```

3.7.3 Compare the Group Categories Using the Adonis Test (See Note 22)

```
>CH.cond<-metadata[rownames(as.matrix((rar.bray.dist))),2]
```

Perform the ADONIS test

```
>adonis.rar.bray<-adonis( CSS.bray.dist ~ CH.cond )
```

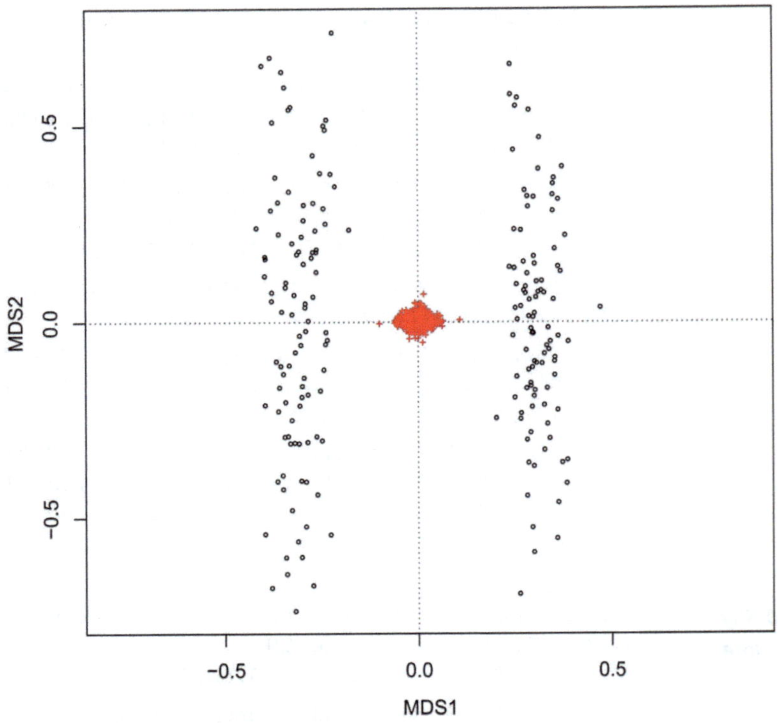

Fig. 3 Default PCoA. The circles represent the samples, the red addition symbols the species. The PCoA was estiamted using the Bray-Curtis index using the CSS normalized dataset. In this plot there is no axis labeling or the adition of color or labes to diferentiate the group samples

Visualize the adonis results

```
>adonis.rar.bray
```

Extract the ADONIS *p*-value

```
>adonis.rar.bray$aov.tab[ 6]$Pr[1]
```

3.7.4 Compare the Variance Homogeneity Between Groups (See Note 23)

Perform the betadisper test

```
>betadisper.res<-betadisper(CSS.bray.dist,
CH.cond, type = c("median"))
```

Look for the significance for the test

```
>permutest(betadisper.res, pairwise = FALSE, per-
mutations = 9999)
```

3.7.5 Represent the PCoA Using Different Function Approximations (See Note 24 and Fig. 4)

Function that decouples data from a vegan function

```
>PCoA.scores<-scores(PCoA)
```

Calculate the eigenvalues

```
>eig <- eigenvals(PCoA)
```

Estimate the proportion of variance explained

```
>eigVariance<- round(eig / sum(eig) * 100,digits=3)
```

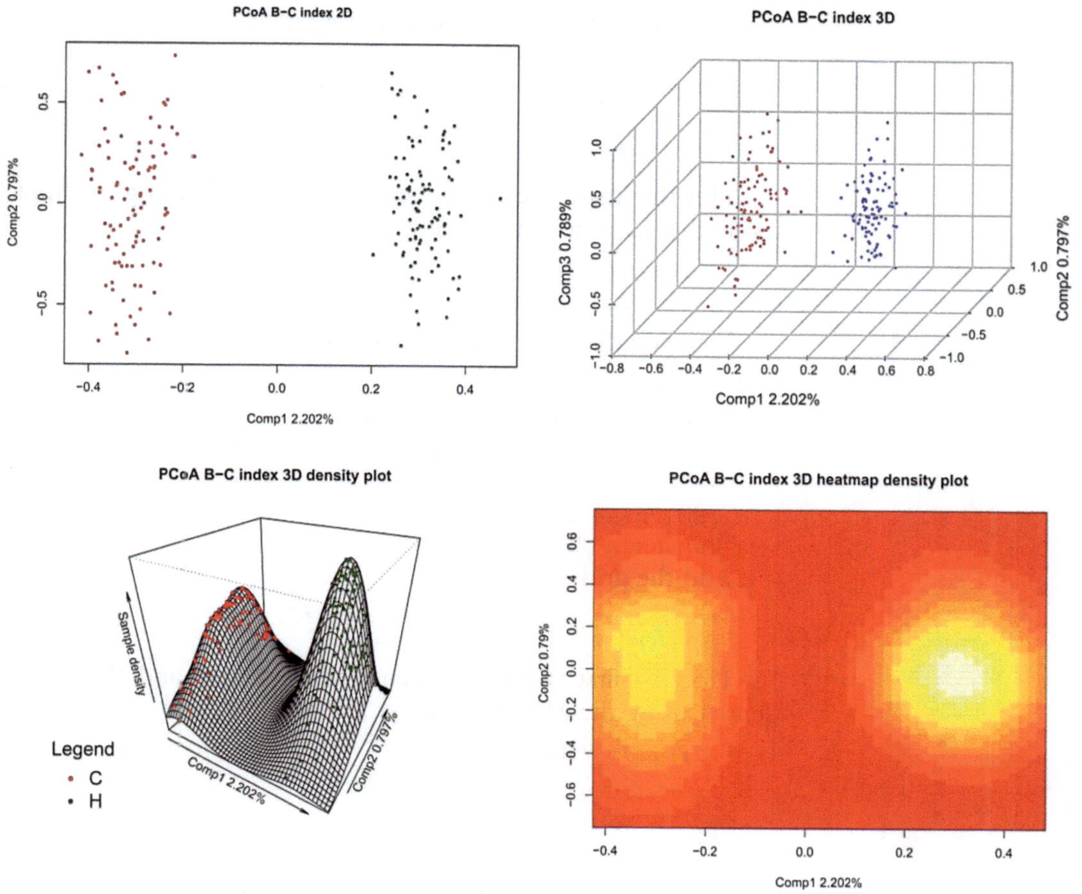

Fig. 4 Different PCoA representations. The top PCoA are based only on the representation of the first two and three components. The lower PCoA representations are based on the projection of the first two components and the sampling density surface. This surface is a continuous model of the distribution of the samples across the first two components and can be visualized as a density plot or as a heatmap. It is useful for the visualization group clustering and dispersion. The PCoA was estimated by the Bray-Curtis index using the CSS normalized dataset

Proportion of variance explained of the first component

```
>x.lab<-paste("Comp1",paste(eigVariance[1],"%",s
ep=""))
```

Proportion of variance explained of the second component

```
>y.lab<-paste("Comp2",paste(eigVariance[2],"%",s
ep=""))
```

Proportion of variance explained of the third component

```
>z.lab<-paste("Comp3",paste(eigVariance[3],"%",s
ep=""))
```

Divide the plot window into 2 rows and 2 columns

```
>pdf("Fig4_Different_PCoA_representations.
pdf",width=12, height=10)
>par(mfrow=c(2,2))
```

Plot the first two components of the PCoA

```
> plot(PCoA.scores$sites, cex = 1, col= c(rep("
red",100),rep("darkgreen",100)), pch = 16, cex.
main=1,cex.sub=0.7, xlab=x.lab, ylab=y.lab,
main="PCoA B-C index 2D")
```

Plot the first three components of the PCoA, using the libraries: FactoClass and scatterplot3d

```
>library("FactoClass")
>library("scatterplot3d")
```

Estimate the scores for the first three components

```
>PCoA.scores<-scores(PCoA,choices=c(1,2,3))
```

Plot the first three components

```
> scatterplot3d(PCoA.scores$sites[, 1:3], pch =
16, grid=FALSE, box=FALSE,color=c(rep("red",100),
rep("blue",100)), xlab=x.lab, ylab=y.lab,zlab=z.
lab,main="PCoA B-C index 3D")
```

Add the grid

```
>addgrids3d(PCoA.scores$sites[, 1:3], grid =
c("xy", "xz", "yz"))
```

Plot the first two components of the PCoA and the sample density using the MASS library

```
>library("MASS")
>x <- PCoA.scores$sites[, 1]
>y <- PCoA.scores$sites[, 2]
>z <- PCoA.scores$sites[, 3]
```

Estimate the kernel density surface (*see* **Note 25**).

```
>dens3d <- kde2d(x, y,n = 50)
> res<-persp(dens3d, box=TRUE, theta=30,phi=30,
xlab=x.lab, ylab=y.lab, zlab="Sample density",
main="PCoA B-C index 3D density plot")
```

Library to plot the sample points into the density plot

```
>library(fields)
```

Create a model to interpolate the values for the first two components into the density surface (*see* **Note 25**)

```
>density.z<- interp.surface( dens3d, cbind(x,y))
>points(trans3d(x, y, density.z, pmat = res), col
= c(rep("red",100),rep("darkgreen",100)), pch =
16, cex=1)
```

Add the plot legend

```
legend("bottomleft", legend=c("C", "H"), pch=16,
col=c("red", "darkgreen"), cex=1.5,box.lty=0,
title="Legend") #
```

Heat density plot

```
>image(dens3d, xlab=x.lab, ylab=y.lab, main="PCoA
B-C index 3D heatmap density plot")
```

Close the plot

```
>dev.off()
```

3.7.6 Comparison of the Efficiency of Different Dissimilarity Matrices (See Note 26 and Fig. 5)

```
>distance_matrix_comparison(viromic.CSS,viromic.rar,metadata) # (see Note 27)
```

Sum variance three comp		ADONIS p-value	betadisper p-value
CSS bray	9.501	0.001	1e-04
CSS hell	26.002	0.001	1e-04
CSS jac	19.515	1.000	1e-04
rar bray	8.998	0.001	1e-04
rar hell	25.821	0.001	1e-04
rar jac	11.811	0.001	1e-04

The results from Fig. 5 and the table show a clear separation between classes C and H. This is independent of the normalization method. It is also observed that the Hellinger distance is the one that greater variation explains for both methods of normalization, while the Jaccard tends to agglomerate all the samples in the same area for the case of the normalization by CSS.

3.7.7 Perform a Nonmetric Multidimensional Scaling Analysis (NMDS) (See Fig. 6)

The Bray-Curtis and the Hellinger distance were the ones that show the best performance. Now we will use these distances to make an NMDS. The NMDS is a nonparametric method to perform an ordination analysis using as input a distance matrix. The NMDS attempt to represent the whole variance of the distance matrix into a predefined number of dimensions (*see* **Note 28**)

```
>pdf("Fig6_NMDS_comparison.pdf",width=17)
>par(mfrow=c(1,3))
```

Create NMDS from using the CSS normalized dataset and the Bray-Curtis index

```
>nmds.CSS.bray<-metaMDS(CSS.bray.dist,
trymax=2000,pc=T) # (see Note 29)
>plot(nmds.CSS.bray$points, cex = 1.5,
col= c(rep("darkgreen",100),rep("r
ed",100)), sub=paste("Stress",round(nmds.CSS.
bray$stress,digits=3),sep="="), cex.main=1.7,cex.
sub=1,pch=16, main=" Non-metric multidimensional
scaling B-C index")
```

Create NMDS from using the CSS normalized dataset and the Hellinger distance

```
>CSS.hell.dist<-vegdist(decostand(viromic.CSS,meth
od="hell"),method="euclidean")
>nmds.CSS.hell<-metaMDS(CSS.hell.dist,
trymax=2000,pc=T)
>plot(nmds.CSS.hell$points, cex = 1.5,
col= c(rep("darkgreen",100),rep("r
ed",100)), sub=paste("Stress",round(nmds.CSS.
hell$stress,digits=3),sep="="), cex.main=1.7,cex.
```

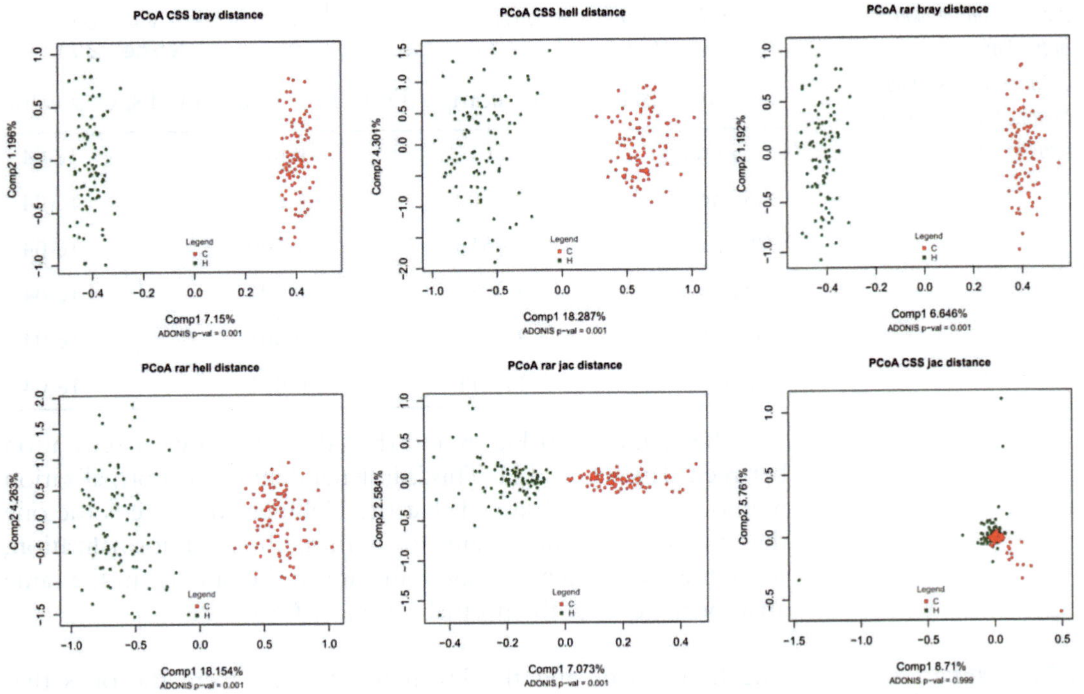

Fig. 5 PCoA using different dissimilarity matrices. Six different PCoA analyses using different distance matrix and normalization methods. Those included the two normalization methods, CSS and rarefaction, and three different dissimilarity indexes: The Bray-Curtis, the Hellinger and the Jaccard. For all the methods the ADONIS group *p*-value was calculated

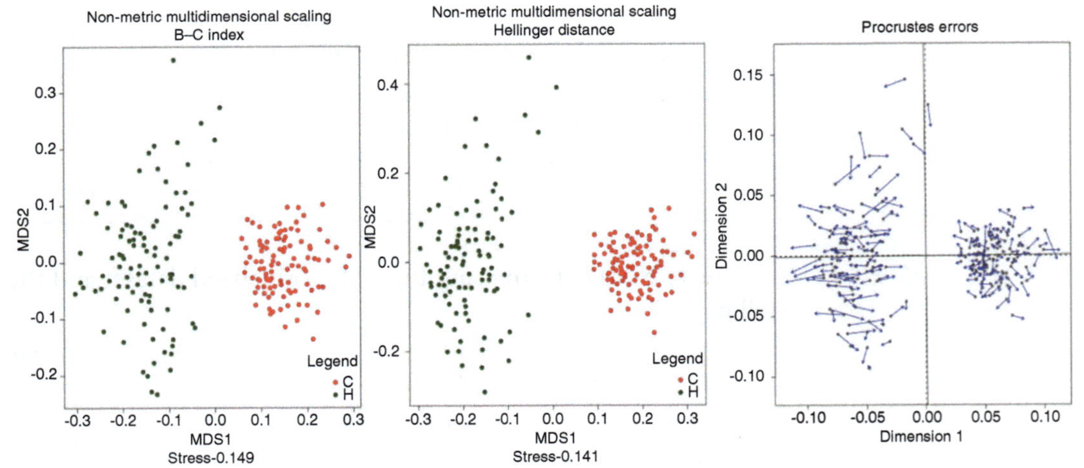

Fig. 6 NMDS comparison. Comparison of two NMDS scaling analyses using the Bray-Curtis and the Hellinger distance. Both NMDS show similar results and present a similar Stress. This indicates that both solutions are suitable for working and that the distribution of the data does not suffer from methodological artifacts. Moreover, the Procrustes analysis shows a statistical congruence between the two-ordination analysis

```
sub=1,pch=16, main=" Non-metric multidimensional
scaling Hellinger distance")
```

3.7.8 Perform a Procrustes Test (See Fig. 6)

To verify if the two ordination analyses are significantly similar we can employ the protest function, this test verifies if the cluster configurations are congruent

```
>prtest<-protest(nmds.CSS.bray,nmds.CSS.hell)
> prtest$signif
>plot(prtest)
```

[1] 0.001

In this case, the *p*-value is below 0.01, this indicates that they are significantly similar.

```
>dev.off()
```

3.8 Clustering Analysis

In Clustering analysis, it is the task of grouping a set of objects in such a way that members of the same group (called a cluster) are more similar, in some sense or another. It is widely used in metagenomic analysis to reveal the structure of subpopulations within environment/condition/groups.

3.8.1 Partitioning Around Medoids (PAM) Algorithm (See Fig. 7)

The PAM algorithm finds a cluster configuration in which the dataset could be divided optimally into K groups, where K is the desired number of groups (for a deeper compression of the algorithm check (https://doi.org/10.1007/s10852-005-9022-1 and https://doi.org/10.1016/j.eswa.2008.01.039)). This clustering method has been used to find the enterotypes [17].

Here, the cluster configuration of the groups, divided from 2 to 20, will be calculated using the PAM algorithm and will be weighted using the silhouette index (*see* **Note 30**)

```
>obs.silhouette.temp=NULL
>nclusters.silhouette=NULL
>pdf("Fig7_NMDS_PAM_clustering.pdf")
```

Estimate the value of the silhouette index for clusters configurations from 2 to 20

```
>num.clusters<-(length(colnames(as.matrix(CSS.
bray.dist)))-180)
>for (k in 1:num.clusters) {
> if (k==1) {
> obs.silhouette.temp[k]=NA
> } else {
> data.cluster.temp<-as.vector(pam(CSS.bray.dist,
k, diss=TRUE)$clustering)
> obs.silhouette.temp<-mean(silhouette(data.clus-
ter.temp, CSS.bray.dist)[,3])
> nclusters.silhouette[k]=obs.silhouette.temp
> }
>}
```

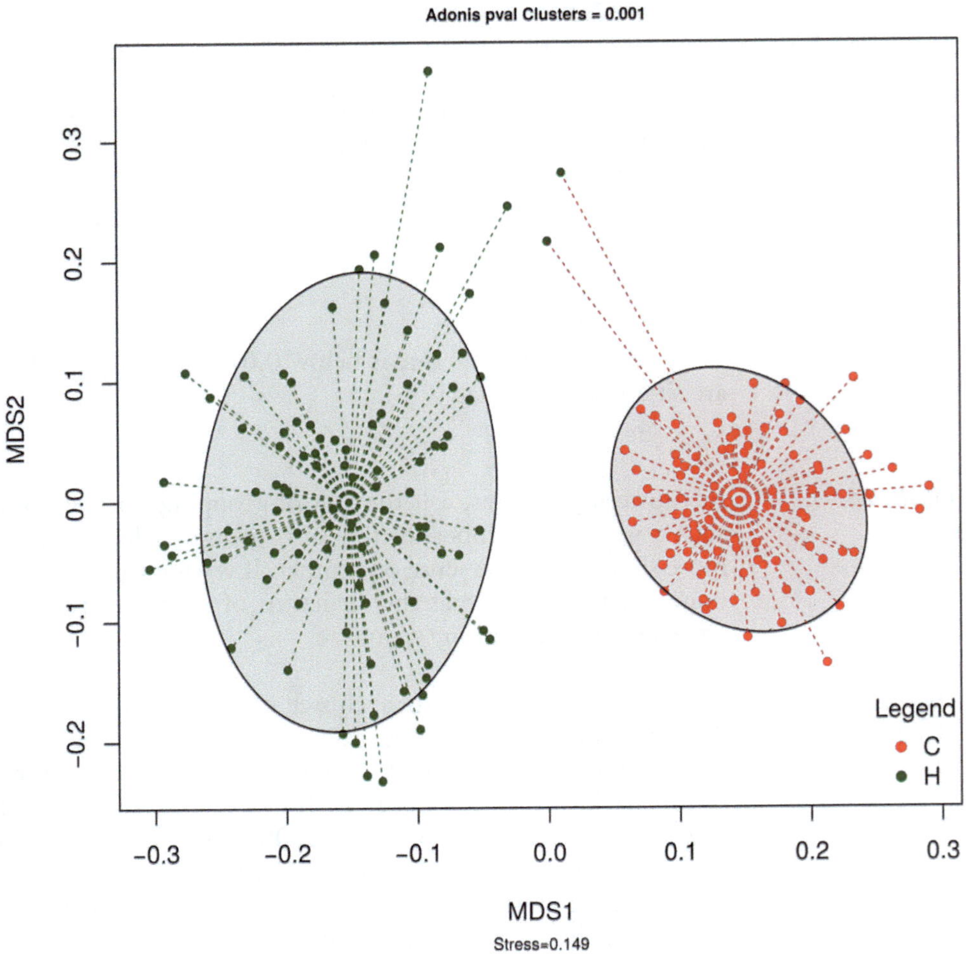

Fig. 7 NMDS PAM clustering. NMDS scaling analysis estimated using the Bray-Curtis index. The ellipses represent the cluster configuration estimated by the PAM algorithm. The ellipses represent the 80% of the group variance. The cluster configuration was the one that maximizes the Silhouette index and was validated by the ADONIS test (p-value = 0.001)

Obtain the cluster configuration that maxims the Silhoutte index

```
>silhouette.num.clusters<-
grep("TRUE",max(nclusters.silhouette, na.rm =
T)==nclusters.silhouette)
```

Perform the cluster configuration

```
data.cluster<-as.vector(pam(CSS.bray.dist, silhou-
ette.num.clusters, diss=TRUE)$clustering)
```

Validate the cluster configuration using the ADONIS test

```
>adonis.clusters<- adonis( CSS.bray.dist ~ data.
cluster )$aov.tab[ 6]$Pr[1]

>clust.title<-paste("Adonis pval
Clusters",round(adonis.clusters,digits=3),sep=" =
")
```

Plot the cluster configuration into the NMDS analysis representing the clusters using the functions "ordiellipse" and "ordispider" available in the vegan library

```
>plot(nmds.CSS.bray$points, cex = 1, pch=20,
col= c(rep("darkgreen",100),rep("red",100)),
main=clust.title, sub=paste("Stress",round(nmds.
CSS.bray$stress,digits=3),sep="="), cex.
main=0.7,cex.sub=0.7)
>treat<-data.cluster
>for(i in unique(treat)) {
> print(i)
> if(i==1){
> color="darkgreen"
> }
> if(i==2){
> color="red"
> }
```

Ellipses confidence is set to the 80%.

```
>ordiellipse(nmds.CSS.bray$point[grep(i,treat),],
groups=treat[treat==i],col=color,draw="polygon",al
pha=40, conf=0.80,show.groups=i)
> ordispider(nmds.CSS.bray$point[grep(i,treat),],
groups=treat[treat==i],col=color,lty=2,lwd=0.7)
>}
>dev.off()
```

3.8.2 Hierarchical Clustering and Heatnaps

The hierarchical clustering seeks to find a hierarchy of clusters. This method is useful in combination with the hetmaps to visualize a detailed cluster configuration and observe the relative abundance of each species in each cluster.

Estimate the hierarchical clustering

```
>hcluster.col <- hclust(CSS.bray.dist, method
="ward.D")
```

Only plot the most abundant species (*see* **Note 31**)

```
>prop.viromic.CSS<-prop.table(viromic.CSS,1) * 100
>apply(prop.viromic.CSS,1,sum)
>vec.above<-apply(prop.viromic.CSS,2,function(x)
{mean(x)>0.5})
>virus.aboce<-names(vec.above[grep(TRUE,vec.
above)])
```

Estimate the row (species) clustering

```
distance.row<-vegdist(t(viromic.CSS[,virus.ab-
oce]))
>hcluster.row <- hclust(distance.row, method
="ward.D")
```

Add the sample condition colors

```
>ColSideColors<-c(rep("red",100),rep("blue",100))
```

Define the colors for the heatmap, function "colorRampPalette" is in the RcolorBrewer library

```
>heatcolors <- colorRampPalette(brewer.pal(9,
"Spectral"))
```

Abbreviate the taxonomic annotation

```
>plot.heatmap.matrix<-t(prop.viromic.CSS)
>rownames(plot.heatmap.matrix)<-
sapply(rownames(plot.heatmap.matrix),function(x)
{unlist(strsplit(x, ";"))[5]})
```

Plot the results

```
>pdf("Fig8_Heatmap_hclust.pdf ") # (see Note 32)
```

Calculate and represent the heatmap, the heatmap.2 function is in the gplot library (Fig. 8)

```
>heatmap.2( plot.heatmap.matrix,
    >#main = paste( "test"),
    >col=heatcolors,
    >trace="none",
    >margins =c(5,7),
    >ColSideColors=ColSideColors,
    >colCol = ColSideColors,
    >dendrogram="both",
    >Rowv = as.dendrogram(hcluster.row),
    >Colv = as.dendrogram(hcluster.col),
    >key.xlab = "% Relative abundance",
    >cexRow =1,
    >cexCol = 0.8,
    >na.rm = TRUE )
>dev.off()
```

3.8.3 Add a Statistical Significance to the Hierarchical Clustering Partitions

The function "pvclust," library pvclust performs a bootstrapping to add statistical significance to the clustering partitions. For more information visit the http://stat.sys.i.kyoto-u.ac.jp/prog/pvclust/ web site (Fig. 9).

```
>pv.hclust <- pvclust(t(viromic.CSS), method.
dist="cor", method.hclust="ward.D",parallel=T,
nboot=1000) # (see Note 33)
>pdf("Fig9_Bootstrapping_HC.pdf",width=24)
>plot(pv.hclust)
```

Highlight with a rectangle those clusters with a significant p-value

```
>pvrect(pv.hclust, alpha=0.95)
>dev.off()
```

3.9 Biomarker Discovery

The determination of different biomarkers (species), which are related to each condition, is not an easy task, the use of wrong

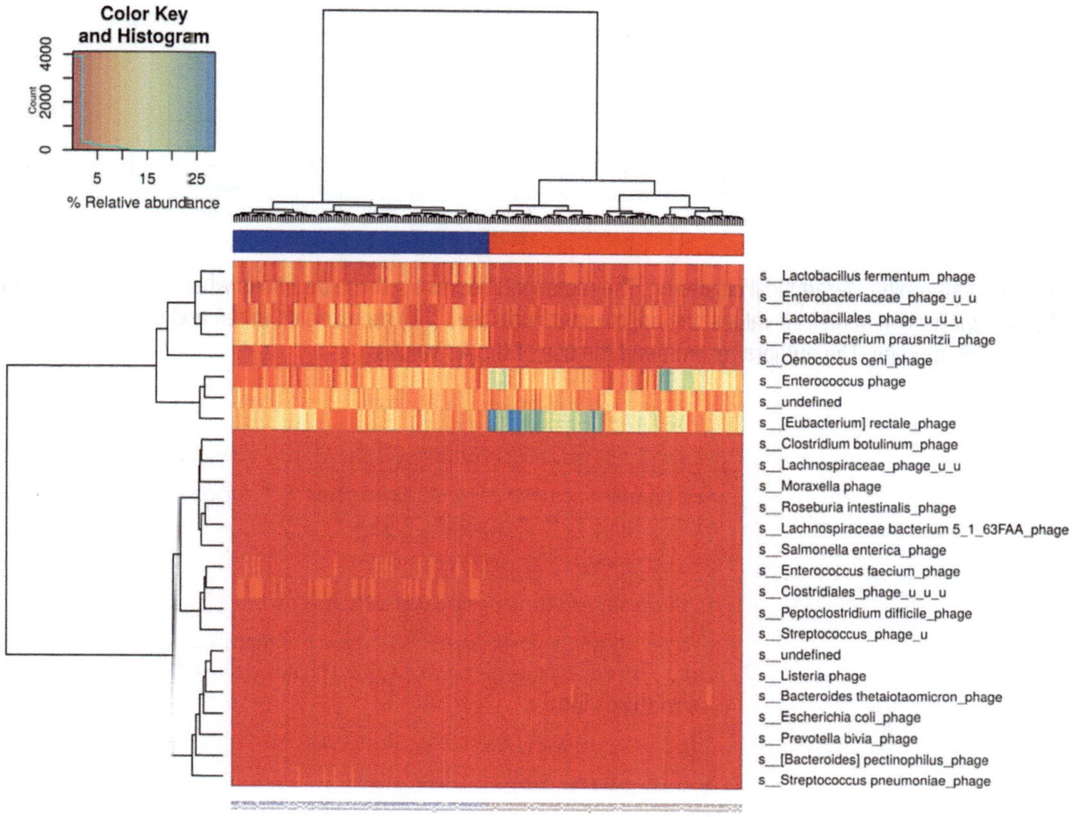

Fig. 8 Heatmap. Hierarchical clustering analysis of the complete dataset represented as the top dendrogram. The heatmap only shows those species whose relative abundance is above 0.5%. The color gradient represents the relative abundance of the spices. The column colors represent the group of the samples

statistical test can generate false positives results or disparage species that may actually be biomarkers. In this section, two statistical packages that have been commonly used in metagenomic studies will be presented. All these use multivariate models and make different corrections to avoid the FDR.

3.9.1 Compute Differential Abundance Analysis Using a Zero-Inflated Log-Normal Model

This approximation is integrated into the metagenoSeq library, function "fitFeatureModel." It is strongly recommended to check the tutorial https://www.bioconductor.org/packages/devel/bioc/vignettes/metagenomeSeq/inst/doc/metagenomeSeq.pdf and the published paper of Paulson et al. [18]. The method fit a zero-inflated log-normal model for each viral specie. Then, perform a log Fold change test between the two conditions and return the p-value and the adjusted p-value. This method can deal with sparsity data which is common in metagenomic viral data.

Create an MRexperiment object (*see* **Note 34**)

```
>Metadata = AnnotatedDataFrame(metadata)

>taxVir<-sapply(rownames(t(viromic.
CSS)),function(x){unlist(strsplit(x, ";"))})
```

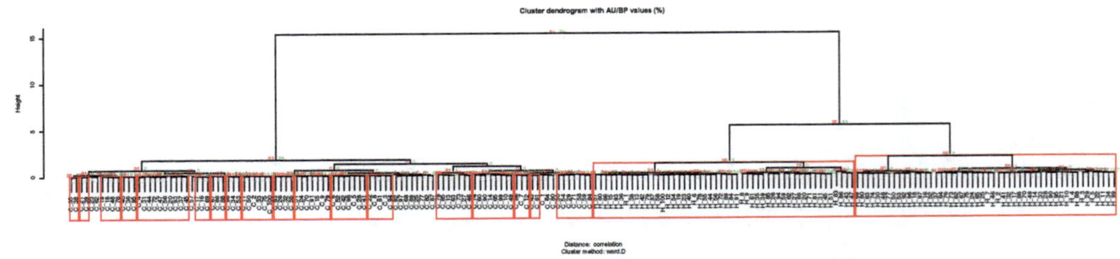

Fig. 9 Bootstrapping hierarchical clustering. The rectangles represent those partitions whose *p*-value is <0.05. The plot shows two bootstrap values, the approximate unbiased (AU) represented in red and the BP (Bootstrap Probability) in green. The authors recommend the use of the AU values

```
>taxVir<-t(taxVir)
>taxVir<-cbind(rownames(taxVir),taxVir)
>colnames(taxVir)=c("Taxonomy","superkingdom","ord
er","family","genus","species")
>taxVir=data.frame(taxVir)
>OTUdata = AnnotatedDataFrame(taxVir)
>obj= newMRexperiment(as.matrix(abundance.table.nr
.sp),phenoData=Metadata,featureData=OTUdata)
```

Normalize data

```
>obj <-cumNorm(obj, p = 0.5)
>pd <-pData(obj)
# Create the model matrix
>mod <-model.matrix(~1 + HC, data = pd)
```

Compute de model using the function fitFeatureModel

```
>vir.mod =fitFeatureModel(obj, mod)
# Extract the results
>df.mod.res<-MRcoefs(vir.mod, number = dim(taxVir)
[1])
 # Select those coefficients which are statisti-
cally significant
>df.mod.sigVirtax<-subset(df.mod.
res,adjPvalues<0.05)
```

Show those specie whose relative abundance is statistically different

```
> df.mod.sigVirtax
```

3.9.2 Linear Discriminant Analysis (LDA) Effect Size (LEfSe) Method

LEfSe is a very popular tool used in metagenomics, it is a three-step algorithm that first perfoms a Kruskal Wallis test between classes, then a Wilcoxon test between all the parwise subclasses and finally a linear discrimiannt analysis. Those species/genes that possess a LDA score above a certain cutoff are selected as biomarkers [19]. LEfSe has an online version in which it details in a very clear and precise way the necessary steps to carry out this analysis https://huttenhower.sph.harvard.edu/galaxy/. It is strongly recommend to use this version unless the server is down, the dataset is very large, or you have to make multiple comparisons.

Create the LEfSe infile in R

```
>LEfSe.infile<-t(viromic.CSS)

>rownames(LEfSe.infile)<-
gsub(";","|",rownames(LEfSe.infile)) # (see Note
36)

>Classes<-as.character(metadata[colnames(LEfSe.
infile),2])

>samples<-colnames(LEfSe.infile)

>LEfSe.infile<-rbind(Classes,samples,LEfSe.infile)

>write.table(LEfSe.infile, "LEfSe.infile" ,col.
names=F,row.names = TRUE,quote=FALSE,sep = "\t")
```

Exit from R

```
q()
```

Perform an analysis with LEfSe (*see* **Note 7**)

```
$LEfSe_dir=~/software/nsegata-lefse-e3cabe93a0d1
```

Define the infiles

```
$ mkdir LEfSe

$ mv LEfSe.infile ./LEfSe

$cd LEfSe

# This is the inflie created in R

$infile=LEfSe.infile

# The current directory

$indir=`pwd`

$outir=$indir/out_lefse_LDA

$mkdir $outir
```

First row contains the class name for each sample "C" or "H"

```
$class=1

# No subclasses

$subclass=0

# Samples

$samples=2
```

Print the help and program options

```
$ python2.7 $LEfSe_dir/format_input.py -h

# format_input.py, create the LEfSe infile

# -o set the normalization value (default -1.0 meaning
no normalization)

$ python2.7 $LEfSe_dir/format_input.py
$indir/$infile $outir/$infile.format_input -c $class
-s $subclass -u $samples -o 1000000
```

Print the help and program options

```
$ python2.7 $LEfSe_dir/run_lefse.py -h
```

Perform the LEfSe analysis

```
# -w set the alpha value for the Wilcoxon test (default 0.05). In this case is going to be 0.00001
# -l set the threshold on the absolute value of the logarithmic LDA score (default 2.0). Change value to 3
# -s 1 set the multiple testing correction options
$ python2.7 $LEfSe_dir/run_lefse.py $outir/$infile.format_input -w 0.00001 -s 1 -l 3 $outir/$infile.res
```

Plot the barplot in a pdf format with a resolution of 600 dpi (*see* **Notes 35** and **36**)

```
$python2.7 $LEfSe_dir/plot_res.py $outir/$infile.res $outir/$infile.pdf --dpi 600 --format pdf
```

Plot the hierarchical cladogram in a pdf format with a resolution of 600 dpi (*see* **Notes 35** and **36**)

```
$ python2.7 $LEfSe_dir/plot_cladogram.py $outir/$infile.res $outir/$infile.cladogram.pdf --dpi 600 --format pdf
```

4 Notes

1. This repository is for Ubuntu 16.04, you should specify your proper Ubuntu R repository. This link https://cran.r-project.org/bin/linux/ubuntu/README.html explains more in detail how to install R.
2. To obtain a better intuition of the R language you can visit the https://www.statmethods.net/r-tutorial/index.html web site.
3. The function "install.packages" will prompt a screen for choosing the country-server from which you would like to download the packages. The installation of the packages can take several minutes depending on the speed of the processor with which you are working. In addition, it may request that programs or libraries be installed in R.
4. The package "metagenomeSeq" belongs to the bioconductor repositories, this is the reason why this package has a different installation.
5. In case that the default python is not python 2.7, you should type the python2.7 before any python script or command, this is to run the scripts under this version. To better understand the pip installation, see this website https://packaging.python.org/tutorials/installing-packages/.
6. To better understand the Anaconda (https://docs.anaconda.com/anaconda/install/) or minconda (https://conda.io/docs/user-guide/install/index.html) setup.

7. There is an easy-to-use web server of LEfSe discovery tool if the number of analyses to be done is small and if the dataset is not very large, this option should be used https://huttenhower.sph.harvard.edu/galaxy/.
8. The directory will contain a sh script (nsegata-lefse-54694b4b0d9e/example/run.sh) which is a clear example of how to use LEfSe.
9. The data come from the simulation of a dataset based on the work of Pérez-Brocal et al. [20]. Two conditions were simulated, the "H" condition which comes from a direct simulation of the real samples and the "C" condition, which comes from a mixture of the real samples and a simulated community which its species abundances follow a power law distribution. About 20 percent of the species were increased in one of two conditions either by adding one random number to each species-sample or by multiplying each species-sample by a random number. The final dataset contains 200 samples and 2299 virus-like species. This can be visualized in R using the command dim() ie:

   >dim(abundance.table)

   [1] 2299 200.
10. The metadata table is created by simply associating each "H" or "C" sample with its corresponding group label.
11. Ordination analyses tend to present problems when the data has a large number of zeros. This type of data is called sparse data and there are statistical methods to deal with it. In the case of viral metagenomes, the high number of zeros can render inoperable even the methods considered to deal with wide data. For this reason, it is recommended to eliminate those viral species that are not present in a certain number of samples. In this case, it will be the 10%.
12. R contains different ways to save the plots in different formats, here it is presented the plot saved in the pdf format, function "pdf" and into the png format using the function "png" this later is commented. This is denoted by adding the "#" symbol before the function. The function dev.off() will close the plot. For this particular example, the rarefaction curve will be saved in the pdf file named Fig1_rarefaction_curves.pdf.
13. Rarefactions curves allow us to quantify the species richness for a given number of samples. These curves are used to determine if the sample sequencing effort is enough to saturate the species diversity. The saturation in the number of species is observed if the rarefaction curve reaches the plateau, this indicates that although the number of reads increases the number of new species remains almost equal. The rarefaction curves

estimation can take several minutes depending on the number of samples and the sequencing effort.

14. The rarefaction to the minimum number of samples is a common technique for data standardization and has obtained good results in simulated datasets [21]. The sample-rarefaction helps to avoid to find statistical biases given the sequencing effort. It is important to that if the sample with the minimum number of reads is very low it is better to remove it from the analysis and take as the reference a sample that reaches the plateau in the rarefactions curves.
15. The Cumulative sum scaling (CSS) normalization is an effective method for sample normalization [18, 20]. The method is based on a quantile normalization for each of the samples of the dataset. The method is specified in the work of Paulson et al. [18] and in the metagenoSeq documentation https://www.bioconductor.org/packages/devel/bioc/vignettes/metagenomeSeq/inst/doc/metagenomeSeq.pdf.
16. All the analyses are based on the rarefied table of species relative-abundances. However, there are tools that are able to estimate the Shannon index based on the spectrum of contigs (see PHACCS [22]). These types of tools have been widely used in the study of viromes and its use is recommended. Since for the present analysis the read data of metagenomes are lacking, this type of approximations cannot be made. The diversity indices used in this analysis were the following:

    Specnumber: is the expected number of species.

    Shannon: quantify how evenly are the species distributed in the sample (The relation between the number of species and the number of individuals in a random sample of animal population).

    Pielou: Quantify the evenness into the species abundances of the community [7].

    Chao1: Richness estimator that adds a correction factor to quantify low abundance species singletons (species observed only once).

    ACE: Similar to the Chao1 index but the correction is for species with fewer than ten individuals.
17. There are several parameters to improve or modify the boxplot, here we are modifying the color of the box ("col" parameter), the axis names ("ylab" and "xlab"), the name of the plot ("main"), and the sub name ("sub"). To see all the modifiable parameters of each function, you can use the command:

    >?Function_name

this will return all the parameters and the description of the function without the need to search the Internet. To exit this help, type q and enter.

18. The beta diversity measures the differences in species composition between habitats. Normally, a dissimilarity index or distance is used to study such differences.

19. The PCoA (also known as Multidimensional scaling) is a method to visualize similarities into a dissimilarity matrix. The method transforms the distance matrix into a set of new coordinates that maximize the variance of the dataset. Then, the set of axes is returned, ordered according to the amount of variance explained. So, the first components are the ones that explain more variance. The vegan function "capscale" is used to calculate a PCoA by first transforming the normalized abundance species-table in a dissimilarity matrix, by default using the Bray-Curtis index. The function needs a formula to perform the analysis, in the case in which informative metadata is available, then it can be incorporated into the formula (see https://www.rdocumentation.org/packages/vegan/versions/1.11-0/topics/capscale). In case you only want to calculate the PCoA of the species-table use this command:

    capscale(data_matrix ~ 1, metaMDS = TRUE,sqrt.dist = TRUE)

20. In ecological analysis, different distances and indexes can be used to perform a PCoA others [8, 9]. The vegan library includes several of these useful distances. In the present example the Bray-Curtis index, the Jaccard distance, and the Hellinger distance are going to be used; this is because they have been used in metagenomic analysis and in viromics [23, 24].

    Bray-Curtis: based to quantify compositional dissimilarity between two samples. It is the ratio of the sum of the lesser values between two species, multiplied by two, divided by the sum of both species.

    Hellinger: the Hellinger distance is the normalization of the data matrix by the Hellinger normalization and then calculate the Euclidean distance between the two samples.

    vegdist(decostand(viromic.CSS,method="hell"),method="euclidean")

    Jaccard: distance estimated based on the presence/absence of the species within the dataset. This distance is thought to be used with binary data (0 and 1).

21. The Jaccard distance is for binary data, then the abundance matrix must be transformed into a presence/absence matrix.

22. The analysis of variance using distance matrices (ADONIS) is a nonparametric Manova suited for data retrieved from dissimilarity index or distances. The statistical power of the test relays into the number of permutations (at least 999). In the present example, the function is used to test if the samples of group H are significantly more distant from those of group C than of themselves. For more information about the ADONIS test, check the [25, 26]. The anosim function also can be used for the same propose http://cc.oulu.fi/~jarioksa/softhelp/vegan/html/anosim.html.
23. The betadisper test if the variance between the groups is similar. If it is not true, this can generate false positives in the ADONIS test, to remedy this, complementary clustering tests can be performed. The significance of the test is given by a permutation test.
24. The PCoA can be improved using different approximations, here the PCoA in 2D, 3D, and the density plot will be represented.
25. A density plot was used to determine the distribution of the samples along the first two components. This method models a 2D density surface based on the sample distribution of the first two components. It can be visualized as a surface ("persp" function) or as a heatmap ("image" function).
26. This function compares different indexes and distances to see which of them explains a greater variation in the data. The comparison includes the distances of Jaccard, Bray-Curtis, and Hellinger. These distances were estimated from the data of the normalized species abundance table and the rarefaction table. The function returns a plot with six different PCoA and a table with the variance explained by the first three components, the ADONIS *p*-value and, the betadisper *p*-value.
27. This function is found in the supplementary library of data "supplementary_diversity_functions". This performs different predetermined comparisons.
28. The NMDS is a powerful statistical method to represent the distance matrix in a low-dimensional space. Before the analysis you have to define the number of dimensions (axis) in which you would like to represent the NMDS (normally two or three dimensions are fine). Then the algorithm tries to find an object configuration that best fits the N-dimensional ordinations space [27] performing multiple iterations that minimize the Stress value. This value represents the differences between the reduced dimensions and the complete dataset. Stress values <0.2 are considered good.
29. The NMDS could be very computing depending on the dataset and sometimes could not reach the optimal solution; for this reason, a fair number of iterations must be done in the case

that the algorithm does not find the optimal solution. In this example, 2000 iterations were done.

30. The Silhouette is a method to validate the consistency of the clusters within the data. The method measures how well each object lies within each cluster.
31. Although the clustering is performed with the complete dataset only the most abundant at species are represented, this is because the output could be very large.
32. Sometimes the output of the function "heatmap.2" library gplots could be truncated; this could be fixed using programs as inkscape (https://inkscape.org/en/). In this program, it is very easy to adjust the size of the drawing canvas.
33. Depending on the number of samples and species the pvclust function can take several hours; it is recommended to use it with the parallel option to parallelize the bootstraps.
34. The MRexperiment object is the data structure that the package metagenoSeq uses for its analyses (check https://www.bioconductor.org/packages/devel/bioc/vignettes/metagenomeSeq/inst/doc/metagenomeSeq.pdf).
35. LEfSe has very nice plotting options, you can specify the full taxonomy of the species separated by "|" ie:

    "k__Viruses|o__Herpesvirales|f__Herpesviridae|g__Cytomegalovirus|s__Cytomegalovirus_u"

    to allow the program to detect which taxonomic level has more influence for sampling segregation.
36. The figures of the LEfSe are in the https://github.com/jorgevazcast/Viromic-diversity/tree/master/LEfSe/out_lefse_LDA.

References

1. Council NR (1999) Perspectives on biodiversity: valuing its role in an everchanging world. The National Academies Press, Washington
2. Whittaker RH (1972) Evolution and measurement of species diversity. Taxon 21:213–251
3. Shannon C (1948) A mathematical theory of communication. Bell Syst Tech J 27:379–423
4. Tuomisto H (2010) A consistent terminology for quantifying species diversity? Yes, it does exist. Oecologia 164:853–860
5. Chao A (1984) Non-parametric estimation of the number of classes in a population. Scand J Stat 1:265–270
6. Chao A, Lee SM (1992) Estimating the number of classes via sample coverage. J Am Stat Assoc:210–217
7. Mulder CPH, Bazeley-White E, Dimitrakopoulos PG et al (2004) Species evenness and productivity in experimental plant communities. Oikos 107:50–63
8. Whittaker RH (1960) Vegetation of the siskiyou mountains, Oregon and California. Ecol Monogr 30:279–338
9. Faith DP, Minchin PR, Belbin L (1987) Compositional dissimilarity as a robust measure of ecological distance. Vegetatio 69:57–68
10. Caporaso J, Kuczynski J, Stombaugh J et al (2010) QIIME allows analysis of high-throughput community sequencing data. Nat Methods 7:335–336
11. Schloss PD, Westcott SL, Ryabin T et al (2009) Introducing mothur: open-source, platform-independent, community-supported software for describing and comparing microbial communities. Appl Environ Microbiol 75:7537–7541

12. Philosof A, Yutin N, Flores-Uribe J et al (2017) Novel abundant oceanic viruses of uncultured marine group II Euryarchaeota. Curr Biol 27:1362–1368
13. Paez-Espino D, Eloe-Fadrosh EA, Pavlopoulos GA et al (2016) Uncovering earth's virome. Nature 536:425–430
14. Vázquez-Castellanos JF, García-López R, Pérez-Brocal V et al (2014) Comparison of different assembly and annotation tools on analysis of simulated viral metagenomic communities in the gut. BMC Genomics 15:37
15. Aitchison J (1981) A new approach to null correlations of proportions. Math Geol 12:175–189
16. Li H (2015) Microbiome, metagenomics, and high-dimensional compositional data analysis. Annu Rev Stat Its Appl 2:73–94
17. Arumugam M, Raes J, Pelletier E et al (2011) Enterotypes of the human gut microbiome. Nature 473:174–180
18. Paulson JN, Stine OC, Bravo HC et al (2013) Differential abundance analysis for microbial marker-gene surveys. Nat Methods 10:1200–1202
19. Segata N, Izard J, Waldron L et al (2011) Metagenomic biomarker discovery and explanation. Genome Biol 12:R60
20. Pérez-Brocal V, García-López R, Nos P et al (2015) Metagenomic analysis of crohn's disease patients identifies changes in the virome and microbiome related to disease status and therapy, and detects potential interactions and biomarkers. Inflamm Bowel Dis 21(11):2515–2532
21. Weiss S, Xu ZZ, Peddada S et al (2017) Normalization and microbial differential abundance strategies depend upon data characteristics. Microbiome 5:27
22. Angly F, Rodriguez-Brito B, Bangor D et al (2005) PHACCS, an online tool for estimating the structure and diversity of uncultured viral communities using metagenomic information. BMC Bioinfo 6:41
23. Reyes A, Haynes M, Hanson N et al (2010) Viruses in the faecal microbiota of monozygotic twins and their mothers. Nature. Nat Publ Group 466:334–338
24. Yatsunenko T, Rey FE, Manary MJ et al (2012) Human gut microbiome viewed across age and geography. Nature 486:222–227
25. Oksanen J, Kindt R, Legendre P et al (2008) Vegan: community ecology package
26. Anderson MJ (2001) A new method for nonparametric multivariate analysis of variance. Austral Ecol 26:32–46
27. Kruskal JB (1964) Multidimensional scaling by optimizing goodness of fit to a nonmetric hypothesis. Psychometrika 29:1–27

Chapter 16

Construction of a Comprehensive Database from the Existing Viral Sequences Available from the International Nucleotide Sequence Database Collaboration

Rodrigo García-López

Abstract

The progress in viromics research has led to the accumulation of a large number of sequences from different types of viruses obtained from different sources. Most databases are specific to different of species or types of viruses. However, raw sequences, as deposited in the reliable online collections, provide a valuable asset in the exploration of genomic and metagenomics datasets.

The International Nucleotide Sequence Database Collaboration (INSDC) is the largest coordinated effort for compiling, sharing, and maintaining the most comprehensive collections of nucleic acids deposited throughout the most important public databases. The compendium includes different types of data such as complete genomes, genes, expressed sequence tags, and data generated by whole genome shotgun analyses spanning all domains of life, as well as the most complete collection of viral sequences available online.

This chapter presents simplified computational methods for the automation of viral nucleotide sequence retrieval from online repositories of the INSDC databases, including all available sequences, except synthetic ones. The subsequent steps can be used for obtaining the taxonomy (including ranks: virus type, baltimore classification, order, family, subfamily, genus and species), and split the database into species subsets to dereplicate the sequences for other downstream applications. Only basic computational knowledge is required.

Key words Computational methods, Databases, Viruses, Genomes, Genes, Taxonomy

1 Introduction

Viromics is still a growing field that has been traditionally limited by reference sequence availability [1]. Although the collections of deposited viral sequences have increased exponentially in recent years (mainly due to the advent of metagenomics) there is still a large gap in the knowledge and exploration of whole viromic datasets, mostly due to the limited pool of reference viral sequences available [2]. Historically, studies have focused on specific species,

Andrés Moya and Vicente Pérez Brocal (eds.), *The Human Virome: Methods and Protocols*, Methods in Molecular Biology, vol. 1838, https://doi.org/10.1007/978-1-4939-8682-8_16,

most often phages affecting model organisms, human health or human-related commercial activities (crops and livestock), which makes up a large percentage of viral contents in public databases [3, 4]. Still, the ecological relevance of environmental viruses has been highlighted as viruses seem to play a relevant role in any biome [5]. This has promoted a deeper and thorough exploration of the viromic contents using whole genomic sequencing approaches, further emphasizing the need for larger and more inclusive viral databases.

Furthermore, viral taxonomy is commonly hindered by a classification system that is often host or pathology-related [6] as viruses share no universal marker and are rather a polyphyletic group [7, 8]. Their taxonomy is further complicated by the large prevalence of horizontal gene transfer since many act as mobile genomic elements that integrate into genomes, further increasing their genetic plasticity [9, 10]. As the naming authority, the International Committee on Taxonomy of Viruses (ICTV) has tackled the complex endeavor of revising classification but its rigid system has not coped well with the advent of high-throughput data from metagenomic studies [6–11].

Different tools have been created for improving viral classification but reference sequences continue to pose as one of the most important features for identification, whether for direct sequence search, or downstream construction of alignments and Hidden Markov Models [12, 13]. As a consequence, there are a number of nucleotide databases that are publicly available, some of which are manually curated with annotations, others just storing collections of genes or genomes [14, 15]. As vast as the number of applications for the sequences is, database selection is a critical process for viromic studies and, for all general purposes, having a comprehensive collection is most convenient.

The largest of its kind is the viral fraction of the nucleotide collection of the International Nucleotide Sequence Database Collaboration (INSDC), part of the synchronized effort of the European Bioinformatics Institute (EBI), DNA Data Bank of Japan (DDBJ), and the National Center for Biotechnology Information (NCBI) to make all their nucleotide sequences available to the general scientific community [16]. For over 30 years, the INSD has been the primary repository for nucleotide sequences that are submitted to any of its partners, coordinating the contents in mirrored images localized in dedicated servers [17], allowing for the preservation and access to data worldwide. Data is constantly exchanged and synchronized across the different servers for updates and fixes.

FTP servers are available for easier sequence access from all three partners, with different features residing in each server. The available datasets from the INSDC include the standard release of the main nucleotide sequence collection, which includes genes and genomic sequences from all domains of life and viruses, as well as expressed sequence tags, genome survey sequences, high-throughput genome sequencing data, transcriptome shotgun assemblies, whole genome shotgun data, and patented items [17]. The identifiers, or accession numbers (shared among all partners in the INSDC) can be used to effectively trace the source organisms or viruses using cross-reference tables pointing to taxonomic identifiers or taxids. These are also available from the FTP servers, enabling the assembly of a complete viral taxonomy for any given sequence and can also be used to separate the sequences based on it, as it is described in the following methods.

2 Materials

2.1 Sequence Retrieval

1. PC with1. Linux 3 or higher (tested on Kubuntu 16.04 distribution).

   Alternatively, macOS or Windows 10 (with Windows subsystem for linux enabled).
2. Bash 4 or equivalent shell.

   As of Sept 2017, default in most current Linux distributions.
3. Over 10 free gigabytes (GB) disk space for storage for all methods (preferable: >20 GB).

   As of Sept 2017 a minimum of 1.2 GB is required for Subheading 2.1.
4. Stable internet connection (>1 Mbps is preferable).

   Current version requires downloading 631 MB of remotely stored data.

   IMPORTANT: All data should be downloaded in the same day to avoid inconsistencies due to updates.

   - Optional: FileZilla 3.7 or higher.

   Installation instructions and files at https://filezilla-project.org/.

2.2 Assembly of Taxonomic Ranks

1. All non-optional items in Subheading 2.1.

   As of Sept 2017, a minimum of 5.3 GB of additional disk space is required for Subheading 2.2.

   Current version requires downloading 4.8 GB of remotely stored data.

   IMPORTANT: All data should be downloaded on the same day to avoid inconsistencies due to updates.
2. Perl v5 of higher.

   As of Sept 2017, included in most current Linux distributions.

- Optional: File containing a list of Accession Numbers (one identifier per line)
- Optional: FileZilla 3.7 or higher.

Installation instructions and files at https://filezilla-project.org/

2.3 Database Split by Species and Dereplication

1. All non-optional items in Subheading 2.2.

   As of Sept 2017, a minimum of 4.2 GB of additional disk space is required for Subheading 2.3.
2. VSEARCH (distributed under GNU General Public License version 3; tested version: 2.4.4).

3 Methods

3.1 Accessing and Retrieving the Remote Files from the INSDC

1. Create a folder for allocating the database and access it (Important: hereafter, the prompt is marked with the initial $ sign and is not part of the command; *see* **Note 1**):

```
$ wdir = ~/VirusDB_2017-09-18/;mkdir $wdir;cd $wdir
```

2. Automatic unsupervised download of all sequences (*see* **Note 2**):

```
$ for set in est gss htg std tsa; do for
type in vrl phg;do for stat in "release/
em_rel" "new/em_cum";do while [ ! -f "01_
INSD/${stat#*/}_${set}_$type.gz" ]; do echo
"Attempting download of ${stat#*/}_${set}_$type.
gz";wget -t 3 -T 2 --user = anonymous --no-
directories --directory-prefix = 01_INSD
ftp://ftp.ebi.ac.uk/pub/databases/fastafiles/
embl${stat}_${set}_${type}.gz 2 > temp.log; if
grep -q "ERROR 404" temp.log;then touch 01_
INSD/${stat#*/}_${set}_$type.gz;echo "${stat#*/}
$set $type does not exist (not an error)";else if
grep -q "Read error at byte" temp.log;then rm 01_
INSD/${stat#*/}_${set}_$type.gz; fi; fi; done; done;
done; done; rm temp.log;find ./01_INSD/ -type f
-empty -delete
```

3. Concatenate and fix headers (*see* **Note 3**):

```
$ for i in $(ls -tr 01_INSD);do y = ${i^^};zcat
01_INSD/$i|sed "s/^>\w*:\w*/>${y:3:11}/"|perl -lne
'next if $_ =~ /^$/;if ($_=~/>/){$line=$_;$line
=~ s/^>\K\w+:\w+ /$1/;($id,$rest)=split(/\
s/,$line,2);$rest =~ s/[\;\:]/ /g;$rest =~ s/[\[\
{]/\(/g; $rest =~ s/[\}\]]/\)/g;$rest =~ s/[^a-zA--
Z\d\s\(\)\.\-\_\/]//g;$rest =~ s/ +/ /g;$rest =~ s/
\.$//;$rest=substr($rest,0,200);print "$id $rest";}
else {print uc($_)}'|gzip >>INSD.gz;done
```

- [Optional: remove the source datasets (no longer required)]

```
rm. -r 01_INSD
```

3.2 Taxonomy

1. Create a folder for the taxonomy:
   $ *mkdir 02_Taxonomy*
2. Create a list of accession numbers from the database in a new folder (*see* **Note 4**):
   $ *zgrep ">" INSD.gz|cut -d " " -f 2|cut -d. -f 1|sort|uniq >02_Taxonomy/INSD.accs*
3. Get the official files for taxonomy from the NCBI taxonomy database (*see* **Note 5**):

```
$ while [ ! -f "02_Taxonomy/taxdump.tar.gz" ];do
echo "Attempting download of taxdump.tar.gz";wget
-t 3 -T 2 --user=anonymous --no-directories
--directory-prefix=02_Taxonomy ftp://ftp.ncbi.nlm.
nih.gov/pub/taxonomy/taxdump.tar.gz 2>taxonomy_
download.log; if grep -q "Read error at byte" tax-
onomy_download.log;then rm 02_Taxonomy/taxdump.tar.
gz;fi;done;rm taxonomy_download.log
```

4. Unpack only the usable files from the tar file (*see* **Note 6**)

```
$ for i in nodes.dmp names.dmp; do tar -C 02_
Taxonomy/ -xzf 02_Taxonomy/taxdump.tar.gz $i;done.
```

5. Unsupervised retrieval of the accession number to taxid cross-reference tables (acc2taxid) from the NCBI (*see* **Note 7**).

```
$ for i in nucl_gb nucl_wgs nucl_gss nucl_est
dead_nucl dead_wgs;do while [ ! -f "02_Taxonomy/$i.
accession2taxid.gz" ];do echo "Attempting down-
load of $i.accession2taxid.gz";wget -t 3 -T 2
--user=anonymous --no-directories --directory-
prefix=02_Taxonomy ftp://ftp.ncbi.nlm.nih.gov/pub/
taxonomy/accession2taxid/$i.accession2taxid.gz
2>taxonomy_download.log; if grep -q "Read error at
byte" taxonomy_download.log;then rm 02_Taxonomy/$i.
accession2taxid.gz;fi;done;done;rm taxonomy_down-
load.log
```

6. Obtain the proprietary scripts for processing the taxonomy (*see* **Note 8**).

```
$ wget https://github.com/rodrigogarlop/VirusDB_
Scripts/raw/master/get_taxid_from_accnum_list.
pl; wget https://github.com/rodrigogarlop/
VirusDB_Scripts/raw/master/get_taxonomy_and_sptax-
id_from_acc2taxid.pl; wget https://github.com/rod-
rigogarlop/VirusDB_Scripts/raw/master/filter_pretax-
onomy_general.pl
```

7. Use the acc2taxid cross-reference tables to obtain terminal node taxids (*see* **Note 9**).

```
$ perl get_taxid_from_accnum_list.pl 02_Taxonomy/
INSD.accs <(cat 02_Taxonomy/*.gz)
```

8. If any taxids are missing, retrieve additional taxids from the corresponding accession numbers using direct requests using the eutils tools from the NCBI servers (*see* **Note 10**). Also, create a list of known bad assignations.

```
$ if [ f 02_Taxonomy/INSD.accs.accs_with_no_tax-
id.txt ];then echo "Processing additional miss-
ing Accession numbers";while read i;do j=$(curl -s
"https://eutils.ncbi.nlm.nih.gov/entrez/eutils/
efetch.fcgi?db=nuccore&id=$i&retmode=text&retty
pe=native"|egrep -A 2 "taxon"|egrep "id \w"|cut
-d "d" -f 2|cut -d " " -f 2|head -n 1) ; printf
"%s\t%s\t%s\t%s\n" $i $i $j $j;done <02_Taxonomy/
INSD.accs.accs_with_no_taxid.txt >02_Taxonomy/last_
chance.txt; awk 'BEGIN {FS="\t"} $4!="" {print}'
02_Taxonomy/last_chance.txt >>INSD.accs.acc_tax-
id.txt;awk 'BEGIN {FS="\t"} $4=="" {print $1}'
02_Taxonomy/last_chance.txt >02_Taxonomy/failed.
txt;fi;printf "0\t10407\n1843481\t1299908\n1843482\
t1299909\n423445\t1891764\n1436891\t2038729\n" >02_
Taxonomy/fix_taxids.txt
```

9. Fix known bad assignations (may be edited accordingly; *see* **Note 11**).

```
$ perl -lane 'BEGIN {open(IN, "02_Taxonomy/
fix_taxids.txt")||die "Cannot open fix_taxids.
txt";while (<IN>){chomp $_;@line = split(/\
t/,$_);$fix{$line[0]}=$line[1] if $line[1] ne
""}}if(exists $fix{$F[2]}) {print "$F[0]\t$F[1]\
t$fix{$F[2]}\t$F[3]"}else{print $_};' 02_Taxonomy/
INSD.accs.acc_taxid.txt >02_Taxonomy/acc2taxid_All_
VirusDB.fix
```

10. Assemble the 22-category taxonomy using terminal-node taxids and create a table with the species-level taxids (*see* **Note 12**).

```
$ rm -f 02_Taxonomy/acc2taxid_All_VirusDB.fix.
bad;perl get_taxonomy_and_sptaxid_from_acc2taxid.
pl 02_Taxonomy/acc2taxid_All_VirusDB.fix;mv 02_
Taxonomy/acc2taxid_All_VirusDB.fix.sptaxid INSD.
sptaxid
```

11. [Important Optional Step: fix outdated taxids] The creation of file *acc2taxid_All_VirusDB.fix.bad* means some node information must be updated in the cross-reference tables. This can be fixed semi-automatically with the following commands followed by repeating commands from **step 9** onward. If this is not corrected, the taxonomy for the corresponding records will be void of taxonomic information and will thus be presented as unidentified sequences (*see* **Note 13**):

```
$ for taxid in $(cat 02_Taxonomy/acc2taxid_All_VirusDB.fix.bad|cut -d " " -f 6);do i=$(grep -m 1 -w "$taxid" 02_Taxonomy/acc2taxid_All_VirusDB.fix|cut -f 1); j=$(curl -s "https://eutils.ncbi.nlm.nih.gov/entrez/eutils/efetch.fcgi?db=nuccore&id=$i&retmode=text&rettype=native"|egrep -A 2 "taxon"|egrep "id \w"|cut -d "d" -f 2|cut -d " " -f 2|head -n 1);printf "$taxid\t$j\n";done >>02_Taxonomy/fix_taxids.txt
```

12. Retrieve viral ranks and append Baltimore classifications and Virus Type for a 7-categories taxonomy (*see* **Note 14**).

```
$ perl 02_Taxonomy/filter_pretaxonomy_general.pl 02_Taxonomy/acc2taxid_All_VirusDB.fix.pretaxonomy >INSD.taxonomy.
```

13. [Optional: remove the taxonomy tables and intermediate steps (no longer required)].

```
$ rm -r 02_Taxonomy
```

3.3 Create Separate Datasets for Each Species and Reduce Redundancy

1. Obtain the proprietary scripts for separating the database (*see* **Note 15**).

```
$ wget https://github.com/rodrigogarlop/VirusDB_Scripts/raw/master/split_fastaDB_by_taxid.pl
```

2. Using the compressed sequence file (from **step 3** in Subheading 3.1) and the species taxonomy file (from **step 11** in Subheading 3.2), create separate sequence files organized by species taxid (*see* **Note 16**).

```
$ perl split_fastaDB_by_taxid.pl INSD.sptaxid INSD.gz
```

3. Define the pathway where VSEARCH is installed (IMPORTANT: change accordingly; *see* **Note 17**):

```
$ vsearch=/path/to/vsearch
```

4. Create a report containing the species taxid, the number of sequences, the total bp, and the taxonomy for each species (*see* **Note 18**).

```
$ printf "#Sptaxid\tSeqs\tSize\tbp\tType\tBaltimore\tOrder\tFamily\tSubfamily\tGenus\tSpecies\n" >Raw_composition.txt ;for i in $(ls -Srh 03_BuildDB/|cut -d. -f 1);do seqs=$(grep -c "^>" 03_BuildDB/$i.fasta|tr -d '\012\015');size=$(ls -lh 03_BuildDB/$i.fasta|sed 's/\s/ /g' |sed 's/ /\t/g'|cut -f 5);name=$(grep -wm 1 "^$i" INSD.taxonomy);bp=$(grep -v "^>" 03_BuildDB/$i.fasta|wc|awk '{print $3-$1}');printf "$i\t$seqs\t$size\t$bp\t$name\n";done >Raw_composition.txt
```

5. Reduce redundancy by dereplicating the datasets and summarize results (*see* **Note 19**):

```
$ rm -f derep_INSD.fasta;printf "#Sptaxid\tSeqs\tSize\tbp\tType\tBaltimore\tOrder\tFamily\tSubfamily\tGenus\tSpecies\n" >Derep_composition.txt;for i in $(ls 03_BuildDB/|cut -d. -f 1);do seqs=$(grep -c "^>" 03_BuildDB/$i.fasta);if [ $seqs -eq 1 ];then cat 03_BuildDB/$i.fasta >03_BuildDB/$i.derep.fasta;else $vsearch --derep_fulllength 03_BuildDB/$i.fasta --fasta_width 80 --output 03_BuildDB/$i.derep.fasta --maxseqlength 1000000000000 --notrunclabels;fi;cat 03_BuildDB/$i.derep.fasta >>INSD_derep.fasta;seqs=$(grep -c "^>" 03_BuildDB/$i.derep.fasta|tr -d '\012\015');size=$(ls -lh 03_BuildDB/$i.derep.fasta|sed 's/\s/ /g'|sed 's/ /\t/g'|cut -f 5);name=$(grep -wm 1 "^$i" INSD.taxonomy|cut -f 2-);bp=$(grep -v "^>" 03_BuildDB/$i.derep.fasta|wc|awk '{print $3-$1}');printf "$i\t$seqs\t$size\t$bp\t$name\n" >>Derep_composition.txt;rm 03_BuildDB/$i.derep.fasta;done
```

4 Notes

1. All computational methods presented here should work by inputting them directly in the command line. The prompt (initial $), should not be copied, it just marks where the command begins.
2. In this method, the sequences from the INSDC are downloaded from the European Bioinformatics Institute as it is readily accessible, it is updated weekly and uses an adequate header format (accession with version and description; since September, 2016, the National Centre for Biotechnology Information uses the same format). The last release is found in ftp.ebi.ac.ukftp.ebi.ac.uk/pub/databases/fastafiles/emblrelease/ and contains the cumulative collection whereas the updates are located at ftp.ebi.ac.ukftp.ebi.ac.uk/pub/databases/fastafiles/emblnew/

   The INSDC virus and bacteriophages files are identified by the vrl and phg suffixes respectively and are contained in the following sets:

   EST: Expressed Sequence Tag

   GSS: Genome Survey Sequence

   HTG: High-Throughput Genome sequencing

   STD: Standard

   TSA: Transcriptome Shotgun Assembly

   PAT: Patent

Patented items are not included in these procedures as the exact sequences are not expected to occur naturally or non-patented variants are preferred.

Please note that not all sets may have updates (a possibility that is considered by the download command). The command will continuously resubmit download requests if the connection is not established correctly unless no file exists in the first place.

As an alternative solution, files may be downloaded using a ftp client such as FileZilla, which is recommended for unstable connections. Please make sure they are located in a folder named *01_INSD* and that the subsequent commands are run from the base folder (*../ 01_INSD;* not from within the *01_INSD* folder).

3. As sequence descriptors in the headers have an unrestricted format, several sequences bear rare characters that must be removed to avoid parsing problems downstream, most notably whenever there is more than one “>“sign (which marks the start of the header/seq duplex). All rare characters are changed to spaces and then collapsed into single spaces. Brackets and the sort are changed to parentheses. Headers longer than 200 characters are truncated as they are also problematic. All sequences are changed to uppercase letters so beware when using masked sequences. Additionally, the name of the subset is added for maximizing the traceability.
4. The assembly of a ranked taxonomy starts with a sequence bearing an accession number as this identifier is linked to a taxonomic identifier (taxid). A single-column file containing these numbers is generated. The version in the accession numbers is ignored to avoid conflicts with outdated items.
5. The taxdump.tar.gz file contains several tools for working out the taxonomy assigned to taxid. It can be downloaded from ftp://ftp.ncbi.nlm.nih.gov/pub/taxonomy/ (optionally with FileZilla).
6. The nodes.dmp is a file containing the links between different taxonomic nodes (from specific branching nodes to the more general root, always in a one-to-one relation) whereas the names.dmp contains the scientific names assigned to each taxid.
7. Cross-reference tables contain sets of accession numbers and their annotated taxids as well as NCBI’s GIs. Although it is possible to download the whole cross-reference tables, it is unadvisable under unstable network conditions. The usage of FileZilla in this step is thus encouraged. To continue, please make sure the files are placed in a folder named *02_Taxonomy* and that the subsequent commands are run from the base

folder (../02_Taxonomy; not from within the *02_Taxonomy* folder).

Protein accession number to taxids cross-reference tables also exist in remote servers.

8. The perl scripts are freely available from github and were created by author Rodrigo García-López and distributed under a GNU 3 license.
9. The script was optimized for low memory consumption and it normally takes ~20–25 min to finish. It stores the loads all accession numbers found in the current downloaded sets and gets those in the taxonomy cross-reference tables where there is a coincidence.

   There is no need to unzip any of the tables.
10. Some acc_nums may have no traceable taxid in the tables but they can be obtained with a remote request to the entrez service via html accession with eutils. This command parses the corresponding remote xml files, which may be slow in some cases but provides updated versions of the accession numbers.
11. Some errors have been reported to exist in the official tables (as downloaded from the NCBI). These must be addressed prior to constructing the taxonomy to avoid bad assignations. This command uses the auxiliary *fix_taxids.txt* created in the previous step, which may be changed according to the missing taxids in the current database (as some may get corrected over time). These types of errors are uncommon but exist mostly permanently in the remote servers.

    IMPORTANT: To avoid some of these inconsistencies, it is advisable to download the sequence data on the same day as the cross-reference tables, since terminal nodes may change when unknown sequences get identified or described, rendering them untraceable.
12. Standardized taxonomies are made of 22 ranked categories, most of which are normally empty for viruses, plus several unranked classifications. Each category is defined by a taxonomic identifier (taxid) which is related to a taxonomic rank (or "no rank"), a scientific name and the link to the next node in the tree (one with higher hierarchy). To assemble the complete taxonomy, the script loads the complete list of links between taxids, the corresponding names and the ranks, and uses the terminal node taxid that is related to an specific accession number (retrieved in previous steps) as the starting point. Each taxid is traced back to its highest level (root of the tree) one node at a time and those bearing ranked categories are organized, effectively constructing a 22 categories taxonomy plus a string containing all unraked categories.

Not all initial taxids point toward the species level but to terminal nodes in the taxonomy, which may be in even lower levels (e.g., serovar/serotype) or just unidentified sequences from known species. The script thus creates an additional list containing the accession number, species-level taxid (sptaxid), and terminal-node taxid (these last two are the same in most cases). The sptaxid is the one used for splitting the database downstream in Subheading 3.3.

This methodology can also be used to process any eukaryotic or prokaryotic sequence, as long as accession numbers and taxids are provided.

13. When taxids get updated (e.g., when unidentified sequences are classified into existing categories) then the terminal taxids change. To avoid this, the nodes.dmp and names.dmp files should be downloaded on the same date as tables from the acc2taxid cross-reference tables (and preferably, on the same date as the sequence data as well). In rare cases, some of them may be outdated. In order to address this, a remote request to the entrez can be carried out with the accession number to retrieve updated taxids. This constructs the input files to append them to the *fix_taxids.txt* file and re-run all commands from **step 9**. Repeat until no *acc2taxid_All_VirusDB.fix.bad* is created in **step 11**, Subheading 3.2.

14. The pretaxtonomy created in **step 10**, Subheading 3.2 contains the 22 ranked categories plus a string of unraked classifications. The script filters them and adds the Baltimore Classification and virus type classification (from unranked categories), additionally dealing with empty categories by including the nearest level that has relevant information for maximum traceability. The output from this step creates the usable taxonomy, containing the related taxid (the one that is found in the sequence headers) and 7 taxonomic categories: Virus type, Baltimore, order, family, subfamily, genus, and species.

15. The perl script is freely available from github and was created by author Rodrigo García-López and distributed under a GNU 3 license.

16. This script is used to organize a large fasta file (.gz) by species based on the accession numbers that are present as the second item in the previously formatted sequence headers (space-separated; e.g., >REL_GSS_VRL AF079492.1 AF079492 Rhesus cytomegalovirus genomic DNA sequence). To do so, it requires the sptaxid crossreference table created in earlier steps which contains three tab-separated columns with accession numbers, species taxids (sptaxids), and terminal-node taxids (commonly matching the sptaxids).

Each newly created fasta file is named after the species taxids contained within (18,621 files in total as of September, 2017).

To do a terminal-node separation instead pass the following alternate command: $ perl split_fastaDB_by_taxid.pl <(cut -f 1,2 INSD.sptaxid) INSD.gz

17. VSEARCH [18] is an open-source alternative to USEARCH [19] suite by Robert C. Edgar that can be used to dereplicate and cluster sequence databases. This step defines the complete pathway of the program's executable file, which depends on the program's installation. Alternatively, USEARCH may be used as the commands are the same (tested version: 8.0).
18. The report table contains the following items for each file (species): taxid, total sequences, used disk space (human readable), total bp, virus type, Baltimore classification, order, family, subfamily, genus, species.
19. By default, VSEARCH discards sequences longer than 50,000 nucleotides. This is overridden by increasing lower limit value with the --maxseqlength option. This approach will swiftly clear off exact identical matches but VSEARCH can also be used to cluster together different sequences with similar identities using the --cluster_fast (memory intensive).

References

1. Beerenwinkel N, Günthard HF, Roth V et al (2012) Challenges and opportunities in estimating viral genetic diversity from next-generation sequencing data. Front Microbiol 3:1–16
2. Pérez-Brocal V, García-López R, Vázquez-Castellanos JF et al (2013) Study of the viral and microbial communities associated with Crohn's disease: a metagenomic approach. Clin Transl Gastroenterol 4:e36
3. Morgan GJ (2016) What is a virus species? Radical pluralism in viral taxonomy. Stud Hist Phil Biol Biomed Sci 59:64–70
4. Calisher CH (2016) The taxonomy of viruses should include viruses. Arch Virol 161:1419–1422
5. Cobián Güemes AG, Youle M, Cantú VA et al (2016) Viruses as winners in the game of life. Annu Rev Virol 3(1):197–214
6. Van Regenmortel MH, Ackermann HW, Calisher CH et al (2013) Virus species polemics: 14 senior virologists oppose a proposed change to the ICTV definition of virus species. Arch Virol 158:1115–1119
7. Edwards R, Rohwer F (2005) Viral metagenomics. Nat Rev Microbiol 3:504–510
8. Proux C, van Sinderen D, Suarez J et al (2002) The dilemma of phage taxonomy illustrated by comparative genomics of Sfi21-like Siphoviridae in lactic acid bacteria. J Bacteriol 184:6026–6036
9. Gibbs AJ (2013) Viral taxonomy needs a spring clean. Its exploration era is over. Virol J 10:254
10. Moreira D, López-García P (2009) Ten reasons to exclude viruses from the tree of life. Nat Rev Microbiol 7:306–311
11. International Committee on Taxonomy of Viruses, ICTV Species List 2016 v 1.2, (https://talk.ictvonline.org/)
12. Roux S, Emerson JB, Eloe-Fadrosh EA et al (2017) Benchmarking viromics: an in silico evaluation of metagenome-enabled estimates of viral community composition and diversity. PeerJ 5:e3817
13. Watkins SC, Putonti C (2017) The use of informativity in the development of robust viromics-based examinations. PeerJ 5:e3281
14. Pickett BE, Sadat EL, Zhang Y et al (2012) ViPR: an open bioinformatics database and analysis resource for virology research. Nucleic Acids Res 40:593–598

15. Liechti R, Gleizes A, Kuznetsov D et al (2010) OpenFluDB, a database for human and animal influenza virus. Database (Oxford) 2010:baq004
16. Cochrane G, Karsch-Mizrachi I, Takagi T (2016) The international nucleotide sequence database collaboration. Nucleic Acids Res 44(D1):D48–D50
17. Nakamura Y, Cochrane G, Karsch-Mizrachi I (2013) The international nucleotide sequence database collaboration. Nucleic Acids Res 41(Database issue):D21–D24
18. Rognes T, Flouri T, Nichols B et al (2016) VSEARCH: a versatile open source tool for metagenomics. PeerJ 4:e2584
19. Edgar RC (2010) Search and clustering orders of magnitude faster than BLAST. Bioinformatics 26(19):2460–2461

Chapter 17

Robust Analysis of Time Series in Virome Metagenomics

Jose Manuel Martí

Abstract

Metagenomics is a powerful tool for assessing the functional and taxonomic contents in biological samples as it makes feasible to study, simultaneously, the whole living community related to a host organism or medium: all the microbes, including virus, bacteria, archaea, fungi, and protists. New DNA and RNA sequencing technologies are dramatically decreasing the cost per sequenced base, so metagenomic sequencing is becoming more and more widespread in biomedical and environmental research. This is opening the possibility of complete longitudinal metagenomic studies, which could unravel the dynamics of microbial communities including intra-microbiome and host-microbiome interactions through in-depth analysis of time series. For viruses, this is particularly interesting because it allows broad interaction studies of viruses and hosts in different time scales, as in bacteria–phages coevolution studies.

This chapter presents computational methods for an automatic and robust analysis of metagenomic time series in virome metagenomics (RATSVM). The same theoretical frame and computational protocol is also suitable for longitudinal studies of spatial series to uncover the dynamics of a microbial community with viruses along a selected dimension in the space. In order to conveniently illustrate the procedure, real data from a published virome study is used. The computational protocol presented here requires only basic computational knowledge. Several scripts have been prepared to ease and automate the most complicate steps, they are available in the RATSVM public repository. For some of the methods a mid-range computing server is advisable, and for some others, it is required. A fat-node with large memory and fast I/O would be the best choice for optimum results.

Key words Robust-analysis, Time-series, Longitudinal metagenomics, RATSVM, Virome, Host-microbiome coevolution, Bacteria-phages coevolution

1 Introduction

Studies of microbial communities by metagenomics are becoming more and more widespread in different biological fields, like in clinical, food, and environmental research [1, 2]. These works are encouraged by new DNA and RNA sequencing technologies that are dramatically decreasing the cost per sequenced base. Thanks to these technological advances, sets of sequences belonging to microbial communities from different sources and times can be

Andrés Moya and Vicente Pérez Brocal (eds.), *The Human Virome: Methods and Protocols*, Methods in Molecular Biology, vol. 1838, https://doi.org/10.1007/978-1-4939-8682-8_17, © Springer Science+Business Media, LLC, part of Springer Nature 2018

retrieved and analyzed to unravel spatial and temporal patterns in the involved microbiota [3].

Generally speaking, biology has recently acquired new conceptual and technological tools to model, investigate, and finally, understand living organisms at a systems level. This has been possible, ultimately, thanks to joint experimental and computational approaches, progress in quantitative techniques, and large-scale measurement methods. Particularly, systems biology strives to unravel the general laws governing the complex behavior of microbial communities [4–6] which includes an attempt for a universal dynamics formulation [7]. Complex microbial communities with viruses and their hosts can be approached in light of ecological theory, which includes general principles like Taylor's law [8, 9] relating the temporal or spatial variability of the population with its statistical average. This law is known to be ubiquitous in the natural world and has been called "fluctuation scale law." It can be found in the core of very diverse systems that could seem unrelated at first sight, such as stock markets [10, 11], random walks [12], animal populations [9, 13, 14], spatial and temporal dynamics in tree forests [15], gene expression in yeasts [16], and even the human genome [17].

Taylor's law has been successfully applied to microbiota spatially [18], with results showing that this population tends to be a concentrated one, far from exhibiting a random or uniform distribution. This ubiquitous law has been tested in experimental settings [19, 20], beyond several efforts to infer the community structure from a dynamic perspective [21–23].

Any rigorous dynamic analysis of the composition of the microbiome requires accurate and reliable tools for assessing the taxonomic contents in the biological samples involved in the study. The best laboratory protocols should be followed by the best computing procedures to minimize any bias from the real composition of the studied microbial community [24]. Post-sampling and cross-over contamination, negative and positive errors, abundance rate distortion, are all examples of subtle and persistent problems that often arise in the practice of metagenomics [25–27]. The challenge of precision in metagenomics can only be tackled by a joint experimental and computational effort in such a direction [24, 28, 29]. On the computational side and in metagenomics, that means, at least, robust procedures able to give a reliable taxonomic composition for each and every sampling point belonging to the longitudinal study. In this way, the last step of the protocol, the time series analysis, is done in the right conditions concerning data quality.

So, one clear goal is to efficiently assign taxonomic labels to the reads down to the species level for reads with reference representation in the used databases and, at the same time, to maintain accuracy in the presence of novel organisms by avoiding overly specific (e.g., species and strain) taxonomic assignments. One important collateral effect of this approach is to alleviate the computational bottleneck by limiting the number of unlabeled reads subjected to additional computational interrogation. The software chosen for accomplishing this central objective has been LMAT [30] (Livermore Metagenomics Analysis Toolkit) from LLNL, with some improvements and the better database available, LMAT-Grand [31].

Nevertheless, before the taxonomic classification, it is recommended to remove all the reads that map against the host, in case it is a single organism not to be included in the study, as happens typically with humans. Even if that is not the case, this step eliminates a very significant amount of human contamination. The classifier could manage these issues to some extent, but from a robust approach, it is quite better to use a specialized method to effectively detect and remove the reads belonging to the host before the step of taxonomic classification. The software selected between several alternatives for an efficient and reliable host removal has been Bowtie2 [32].

An unavoidable set of software and data is provided by NCBI. Particularly useful here is the SRA toolkit software, which takes two subsections in this chapter, and the NCBI SRA databases [33]. In other part of the protocol, they are also indirectly accessed when using entrez [34]. Another essential tool when working with SAM files is the SAMtools [35]. Git is in the center of the management for all the software repositories used along the protocol [36].

Figure 1 summarizes the main stages of RATSVM protocol, distributed in an imaginary axis from pure preparation steps to pure computational ones. With the aim of illustrating the procedure conveniently, real data from a published time series study about the human gut virome is used along the chapter [37].

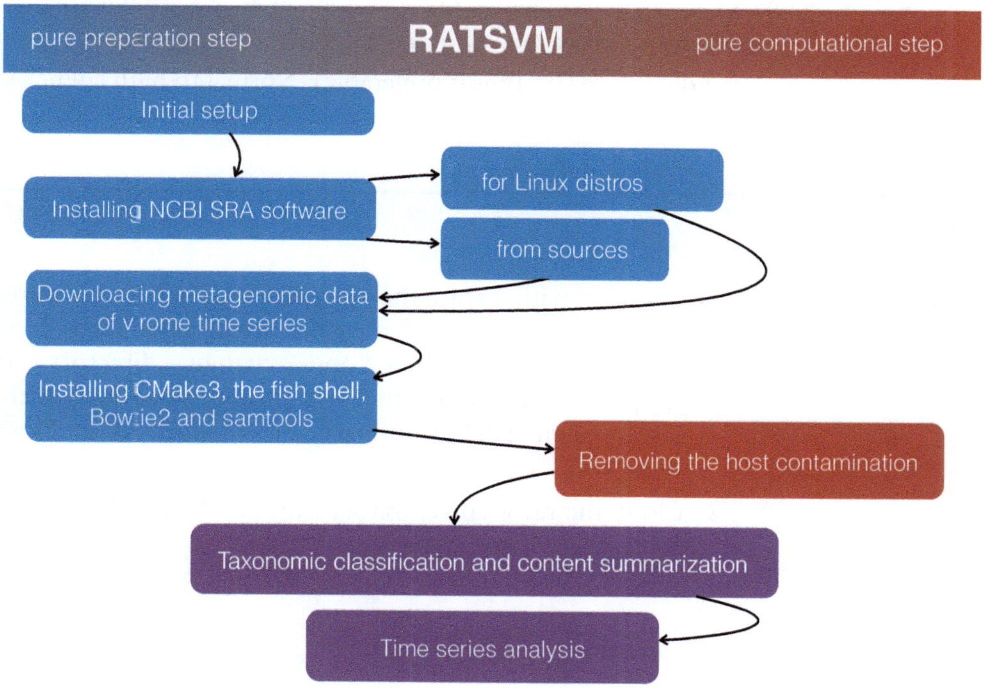

Fig. 1 Main stages of RATSVM protocol, distributed in an imaginary axis from pure preparation steps to pure computational ones

2 Materials

2.1 Overview

This biocomputing protocol, as well as state-of-the-art laboratory protocols concerning viruses, requires advanced equipment. For some of the methods of the chapter a mid-range computing server is advisable, and for some others, it is required. A fat-node with large memory and fast I/O would be the best choice for optimum results.

2.2 Details

Hardware: For best results, it is recommended a fat-node, that is, a SMP (Symmetric MultiProcessing) system with a large quantity of memory per core and a wide-bandwidth and low-latency access to memory and scratch I/O. A memory of ~0.5 TB is required. Intel processors or compatible are required, as some software will be using SSE4 (Streaming SIMD Extensions 4) for increased performance. The total disk space needed to perform the entire protocol is about 1.5 TB in a fast disk. For at least 0.5 TB of them, a local SSD (Solid-State Drive) is advised.

Software: A GNU/Linux distribution is recommended. If the other conditions are met, other flavors of UNIX could be used too, including OS X. A bash 4.2 shell could be used, though the fish shell 2 is recommended (installation instructions included in the protocol). Basic C/C++ development environment, including GNU GCC compilers and essential development libraries, are usually available by default in any GNU/Linux system. Python 3.5 or later is needed. Additionally, for the last step of the protocol the following python packages are required: pandas >1.16.0 with openpyxl, numpy, scipy, matplotlib >1.5.0 and xlrd >0.9. Detailed installation instructions are included in the protocol for any other software.

Network: Stable internet connection with medium to large bandwidth (>10 Mbps is preferable). Normal to high latency levels are acceptable.

3 Methods

3.1 Initial Setup

1. To begin with, create the ratsv directory where software and data will be allocated for Robust Analysis of Time Series in Virome Metagenomics and change directory there (Important: hereafter, the user prompt is marked with the initial > sign which is not part of the command; *see* **Note 1**):
   ```
   > mkdir ~/ratsvm; cd ~/ratsvm
   ```
2. Clone the ratsv git repository in GitHub (*see* **Note 2**) which contains tools and data:
   ```
   > git clone https://github.com/khyox/ratsvm.git
   ```

3.2 Installing NCBI SRA Software for 64 bits CentOS, Ubuntu or MacOS

As of March 2018, NCBI SRA Toolkit (*see* **Note 3**) is available pre-compiled for some operating systems and architectures, like:

- CentOS Linux 64 bits https://ftp-trace.ncbi.nlm.nih.gov/sra/sdk/current/sratoolkit.current-centos_linux64.tar.gz
- Ubuntu Linux 64 bits: https://ftp-trace.ncbi.nlm.nih.gov/sra/sdk/current/sratoolkit.current-ubuntu64.tar.gz
- MacOS 64 bits: https://ftp-trace.ncbi.nlm.nih.gov/sra/sdk/current/sratoolkit.current-mac64.tar.gz

The updated list is available in https://www.ncbi.nlm.nih.gov/sra/docs/toolkitsoft/. If your system is not there, please proceed to the next section. We will continue this section supposing that your system is a 64 bits CentOS Linux or equivalent since the steps are just the same for the mentioned architectures with the difference of the download link and file name:

1. Download and expand the pre-compiled package:
```
> wget https://ftp-trace.ncbi.nlm.nih.gov/sra/sdk/current/sratoolkit.current-centos_linux64.tar.gz
> tar xvzf sratoolkit.current-centos_linux64.tar.gz
```
2. Rename the resulting inflated folder, whose name could depend on the specific release, for instance, alternatively:
```
> mv sratoolkit.current-centos_linux64 sra-tools
> mv sratoolkit.2.9.0-centos_linux64 sra-tools
```
3. Check the fastdump SRA Toolkit binary:
```
> ~/ratsvm/sra-tools/bin/fastq-dump -version
```

The resulting version number should be 2.8.2 or higher. If so, please skip the next section.

3.3 Installing NCBI SRA Software from Sources

1. Create a directory where NCBI software and data will be allocated and move there:
```
> mkdir ~/ratsvm/ncbi_src; cd ~/ratsvm/ncbi_src
```
2. Check that your system has as many of the following optional prerequisites as possible: libraries and development headers for XML2, Magic and HDF5 (*see* **Note 4**).
3. Clone the required NCBI SRA software git repositories from GitHub (*see* **Note 2**):
```
> git clone https://github.com/ncbi/ncbi-vdb.git
> git clone https://github.com/ncbi/ngs.git
> git clone https://github.com/ncbi/sra-tools.git
```
4. Configure the NCBI VDB software (*see* **Note 5**):
```
> cd ncbi-vdb;configure -prefix = ~/ratsvm/ncbi-vdb
```

5. Build the NCBI VDB software:
```
> make
```
6. Install the NCBI VDB software in ~/ratsvm/ncbi-vdb:
```
> make install
```
7. Analogously, configure, build, and install the NGS software (*see* **Note 5**):
```
> cd ~/ratsvm/ncbi_src/ngs; ./configure
-prefix = ~/ratsvm/ngs; make; make install
```
8. Configure, build, and install the SRA Toolkit software (*see* **Note 5**):
```
> cd ~/ratsvm/ncbi_src/sra-tools; ./configure
-prefix = ~/ratsvm/sra-tools; make; make install
```
9. Check the fastq-dump SRA Toolkit binary and confirm that the resulting version number is 2.8.2 or higher:
```
> ~/ratsvm/sra-tools/bin/fastq-dump -version
```

3.4 Downloading Metagenomic Data of Virome Time Series

1. Clone the entrez (*see* **Note 6**) git repository in GitHub (*see* **Note 2**):
```
> git clone https://github.com/jordibc/entrez
```
2. As an example, get the paired-ends set of FASTQ files from all the runs in the study "Rapid evolution of the human gut virome," a longitudinal characterization of intestinal viral particles from WGS (whole genome shotgun) sequencing. The public sequences of this study are already trimmed and quality-checked. The associated NCBI SRA code is SRP021107, so we run the following python scripts (*see* **Note 7**):
```
> entrez/sra2runacc.py -s SRP021107 | rats-
vm/runs2fastq.py -s sra-tools
```
3. Rename files with the script in the ratsvm repo in order to have meaningful information about the sampling (time and replica) in the filename:
```
> ratsvm/rename_fastq.sh
```

3.5 Installing CMake3, the Fish Shell, Bowtie2, and Samtools

1. If CMake3 is not installed in your system (*see* **Note 8**) you can install it locally for your user by cloning the git repository:
```
> git clone https://gitlab.kitware.com/
cmake/cmake.git; cd cmake/
```
2. Configure the CMake3 build for a local installation under ~/ratsvm/CMake directory:
```
> ./bootstrap -prefix =  ~/ratsvm/CMake
```
3. Build and install CMake3 in ~/ratsvm/CMake directory:
```
> make -j 16; make install
```
4. Check the cmake binary and confirm that the installed version is 3.11 or higher, though any release over 3.2 would be acceptable for this protocol:
```
> cd ..; CMake/bin/cmake --version
```

5. If the fish shell is not installed in your system (*see* **Note 9**) you can install it locally for your user by cloning the git repository in GitHub (*see* **Note 2**):
```
> git clone https://github.com/fish-shell/
fish-shell.git; cd fish-shell/
```
6. Use CMake3 to build and install fish locally under ~/ratsvm/fish directory:
```
> mkdir build
> ~/ratsvm/CMake/bin/cmake -DCMAKE_INSTALL_
PREFIX:PATH~/ratsvm/fish -DCMAKE_BUILD_
TYPE=Release..
> make install
```
7. Test the fish shell:
```
> cd ~/ratsvm/; fish/bin/fish
```
8. Exit from the fish shell session with "Ctrl+D" or exit.
9. Install Bowtie2 (*see* **Note 10**) locally for your user by cloning the git repository in GitHub (*see* **Note 2**):
```
> git clone https://github.com/BenLangmead/
bowtie2.git
> cd bowtie2; make -j 16; cd..
```
10. Install HTSlib (*see* **Note 11**) from the GitHub repository (*see* **Note 2**):
```
> git clone https://github.com/samtools/htslib.git
> cd htslib; autoheader; autoconf -Wno-syntax
> ./configure --prefix = ~/ratsvm/HTSlib
> make -j 16; make install; cd..
```
11. Install SAMtools (*see* **Note 11**) from the GitHub repository (*see* **Note 2**):
```
> git clone https://github.com/samtools/
samtools.git
> cd samtools; autoheader; autoconf -Wno-syntax
> ./configure --prefix = ~/ratsvm/SAMtools
> make -j 16; make install; cd..
```

3.6 Removing the Host (Human) Contamination

1. From NCBI RefSeq database, get human mRNA and ncRNA sequences to extend a human DNA database:
```
> mkdir HSS; cd HSS
> wget ftp://ftp.ncbi.nih.gov/refseq/H_
sapiens/H_sapiens/RNA/rna.fa.gz
```
2. Launch the make_hg19plusRNA.sh script to download the UCSC hg19 human genome reference and manage the building of a corresponding bowtie2 index (*see* **Note 12**):
```
> ~/ratsvm/ratsvm/make_hg19plusRNA.sh
```

3. Run the launch_bt2.fish script to manage the alignment of the reads against the extended hg19 human genome reference prepared in the previous step (*see* **Note 13**):
   ```
   > cd ~/ratsvm; ratsvm/launch_bt2.fish --very-sensitive --bt2_path ~/ratsvm/bowtie2 --db_path ~/ratsvm/HSS --proc 32 hg19plusRNA
   ```
4. Execute the launch_samtools.fish script to obtain FASTQ files for the sequences that did not align against the extended hg19 human genome (*see* **Note 14**):
   ```
   > ratsvm/launch_samtools.fish --samtls_path ~/ratsvm/SAMtools --proc 32 hg19plusRNA
   ```
5. Check that we have the expected number of FASTQ files with the unmapped sequences (27 WGS runs, so 27 FASTQ files):
   ```
   > ls -1 *_unmap.fastq | wc -l
   ```
6. Move the FASTQ files with the unmapped:
   ```
   > mkdir unmap; mv *_unmap.fastq unmap/
   ```

3.7 Taxonomic Classification and Content Summarization

1. Install LMAT (Livermore Metagenomics Analysis Toolkit, *see* **Note 15**) starting by cloning the git repository in GitHub (*see* **Note 2**):
   ```
   > git clone https://github.com/LivGen/LMAT.git; cd LMAT/
   ```
2. Use redoall wrapper around CMake3 to easily build LMAT in place using GNU compilers (*see* **Note 16**):
   ```
   > ./redoall.sh R gnu
   ```
3. Download the LMAT runtime inputs:
   ```
   > bin/get_db.sh --dtype=inputs
   --name=04072014 -outdir=.
   ```
4. Set the environment variable LMAT_DIR to the runtime inputs directory downloaded:
   ```
   > set -x LMAT_DIR ~/ratsvm/LMAT/runtime_inputs/
   ```
5. Download and move the LMAT Grand database (*see* **Note 17**):
   ```
   > bin/get_db.sh --dtype=db --name=lmat-4-14.20mer.db --outdir=.
   > mv lmat-4-14.20mer.db /fastdisk; cd ~/ratsvm
   ```
6. Use pyLMAT_rl.py script to automate processing all the samples with the LMAT taxonomic classifier (*see* **Note 18**):
   ```
   > ratsvm/pyLMAT_rl.py --path unmap --bdir ~/ratsvm --fulldbdir /fastdisk --lmat ~/ratsvm/LMAT/
   --threads 32 --minscore 0 --wgs
   ```
7. Run the pyLMAT_cs.py script to automate processing all the samples with the LMAT content summarization step (*see* **Note 19**):
   ```
   > ratsvm/pyLMAT_cs.py
   ```

3.8 Time Series Analysis

1. The first and second steps in this section are for preparing the results of the LMAT taxonomic classifier and content summarization to the format needed by cmplxcruncher, the code that will perform the time series analysis. To begin with, use the rawlmat2lmat.py script, which will show a progress bar (*see* **Note 20**):

```
> mdir virome; ratsvm/rawlmat2lmat.py
```

2. Run lmat2cmplx.py script to convert the LMAT output (*see* **Note 20**):

```
> ratsvm/rawlmat2lmat.py
```

3. Check that the virome directory contains four Excel spreadsheets:

```
> ls virome/LMAT_*.xlsx
```

4. Run cmplxcruncher against the virome directory in automatic mode (*see* **Note 21**):

```
> ratsvm/cmplxcruncher.py --path virome/
--automatic pdf
```

5. The code generates detailed and exhaustive results for the time series analysis in the results subdirectory (*see* **Note 21**):

```
> cd virome/results; ls -R
```

Figure 2 shows an example of the output achieved in Taylor's Law plot, with a log-LR (Linear Regression) model for fitting the power law.

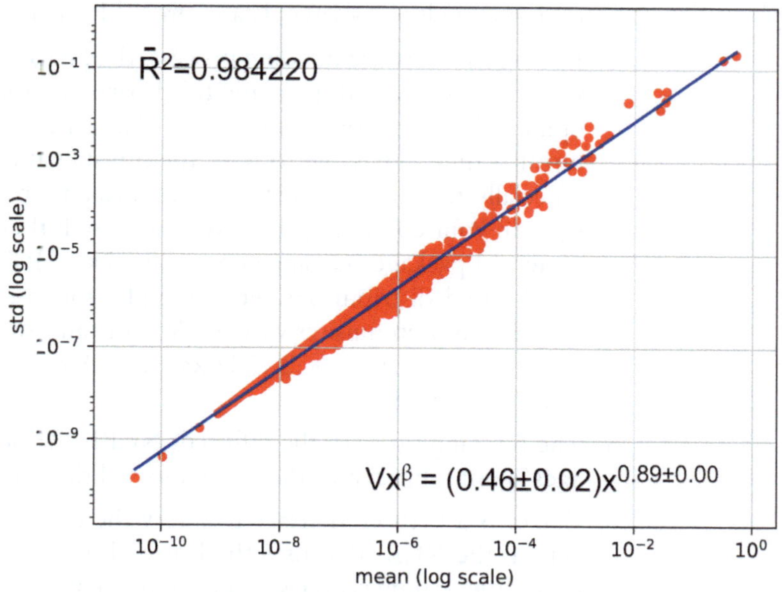

Fig. 2 Taylor's Law plot by fitting a log-L R (Linear Regression) model for the results obtained by LMAT "content summary" step at genus level along the time series

4 Notes

1. All computational methods presented here should work by inputting them directly in the command line. The prompt (initial >) should not be copied, it just marks where the command begins.
2. Git is a Distributed Version Control System (DVCS) that thinks of its data like a stream of snapshots of a mini filesystem, called the repository. Git was founded by Linus Tovals, the creator of the kernel of the GNU/Linux operating system, and it has quickly become the most used DVCS. A condensed course on Git titled "Git in a nutshell," from the author, is available here: http://som1.ific.uv.es/gitdoc. GitHub is the largest web-based hosting service for Git repositories. To interact with GitHub, it is recommended Git version 1.7.10 or higher. The latest version of Git is always available on the Git website: https://git-scm.com/downloads.
3. The Sequence Read Archive (SRA) of the US National Center for Biotechnology Information (NCBI) makes biological sequence data available to the world research community with the goal of enhancing reproducibility and allowing for new discoveries by comparing datasets. The NCBI SRA stores raw sequencing data and alignment information from high-throughput sequencing platforms. It can be manually accessed at https://www.ncbi.nlm.nih.gov/sra/ to search and retrieve sequences of interest. The SRA Toolkit is the recommended method by NCBI of retrieving multiple or large datasets from SRA database.
4. During the software building process, the NCBI SRA software will try to locate and use some third-party libraries and development headers from your system. They are not mandatory but it is quite advisable to have them installed in the system, especially the XML2 library. If you have administrative privileges in your computing box, you can install them using your preferred package-management utility, like APT or YUM for RPM-based distributions. For example, for installing XML2 and HDF5 prerequisites with YUM you can issue:
   ```
   > sudo yum install libxml2 libxml2-devel
   hdf5 hdf5-devel
   ```
5. The building order of the different NCBI SRA software packages is not arbitrary, as they have crossed dependencies.
6. Entrez is a Python interface to query the biological databases kept at the NCBI that uses the Entrez Programming Utilities (E-utilities), nine server-side programs that access the Entrez query and database system at NCBI.
7. The entrez sra2runacc.py script takes in his -s or --sra parameter a SRA database code or identifier (SRP021107 in our

example) and provides a list with the accession codes for all the runs related to the SRA identifier (27 runs in our case). The ratsvm runs2fastq.py script reads this list from the standard input (or from a file using the -f or --file parameter) and manages the retrieval of the FASTQ files using the NCBI SRA-tools and splitting them in different files for each paired-end. If the SRA-tools have been installed in a different location from the place indicated in this procedure, the -s or –sratoolkit parameter of runs2fastq.py should be updated accordingly. The download of any set of FASTQ files related to a NCBI bioproject, including this one, takes some time and requires some available disk space. In our example, the files use more than 310 GB in disk and could take more than 1 h in the download, depending on the network bandwidth. The runs2fastq.py progress indicator will help with this, but please note that it is showing complete downloads, so it will take some time to change from the starting 0% indication and this is totally normal.

8. CMake (https://cmake.org) is a cross-platform, open-source build system generator which eases deployment for software packages, especially for large or complex ones with several dependencies. If you have administrative privileges in your computing box, you can install CMake3 using your preferred package-management utility, like APT or YUM for RPM-based distributions. For example, for YUM you can issue:

```
> sudo yum install cmake3
```

9. The fish shell (https://fishshell.com/) is new-generation smart and user-friendly command line shell that has quite advantages over older shells not only for interactive use but also for programming, like a clean and easy to remember syntax. A good programming guide for the fish shell can be accessed here: http://som1.ific.uv.es/fishprog. In the fish shell website (https://fishshell.com/) there is a download section with links for different operating systems and GNU/Linux distributions. As this is a shell, the last installation step will require administrative privileges. An alternative option, covered in the protocol, is a local installation starting from the git repository and using CMake to build the software. Even with his choice, some basic system prerequisites are needed. If CMake reports that it could not find curses, try installing a curses development package and build again. Depending on if you have, on the one hand, Debian or Ubuntu or, on the other hand, RedHat, CentOS or Amazon EC2, you will issue, respectively:

```
> sudo apt-get install build-essential ncurs-
es-dev libncurses5-dev gettext autoconf
> sudo yum install ncurses-devel
```

10. Bowtie2 (http://bowtie-bio.sf.net/bowtie2) is an ultrafast and memory-efficient tool for aligning sequencing reads to long reference sequences developed at Johns Hopkins University, Baltimore, USA.
11. SAMtools is a set of tools for manipulating next-generation sequencing data, like SAM, BAM, and CRAM files. SAMtools is based on HTSlib, a C library for reading/writing high-throughput sequencing data, which should be installed first. The installation from git repositories described in the protocol is the preferred approach, but if in case of any trouble an alternate installation from packages could be used:
```
> wget https://github.com/samtools/sam-
tools/releases/download/1.7/samtools-
1.7.tar.bz2
> bunzip2 -c samtools-1.7.tar.bz2 | tar xvf -
> ./configure --prefix=~/ratsvm/SAMtools
> make -j 16; make install; cd ..
```
12. The make_hg19plusRNA.sh script downloads the human chromosomes 1 to 22, X, Y and the mtDNA. Those sequences join the mRNA and ncRNA sequences already downloaded in the previous step. Then, ~/ratsvm/bowtie2/bowtie2-build is called in order to build the bowtie2 reference database. This process is time-consuming and could take more than 3 h depending on the network speed and especially on the computing power.
13. The launch_bt2.fish script requires just one positional argument, the name of the bowtie2 alignment database. It also accepts various options whose default values have been chosen to be coherent with the previous steps in this protocol: -b/--bt2_path PATH for the Bowtie2 path, -d/--db_path PATH for the database path, -p/--proc PROC for selecting the number of processors used (of course, the time consumed will be highly dependent on this parameter) and -v/--very-sensitive to enable bowtie2 very sensitive mode (recommended though it will take some more time). The alignment of all the samples against the human reference is CPU-consuming so a computer with more computing cores available will mean a clear advantage. As a reference, using 32 cores, this step could take more than 1 h.
14. As the launch_bt2.fish script (*see* **Note 13**), the launch_samtools.fish script requires the name of the bowtie2 alignment database as positional argument. It also accepts various options with default values chosen to be coherent with the previous steps in this protocol: -s/--samtls_path PATH for the SAMtools package installation path, -p/--proc PROC for selecting the number of processors used in sorting the sequences. Internally, the script will pipe the data through different tools in the SAMtools package, starting with the SAM

files generated by bowtie2 in the previous step until obtaining the FASTQ files for the unmapped sequences. Again, this process takes some time, especially the intermediate stage of sorting the sequences, but not as much as the previous step.

15. LMAT (Livermore Metagenomics Analysis Toolkit) is an all-in-1 metagenomic analysis toolkit including taxonomic classification, content summarization, and gene identification, by Lawrence Livermore National Laboratory. LLNL's history of always having one of the most power supercomputing systems in the world has left its mark in LMAT as it is a very powerful software but can be very demanding in computing resources. In fact, for using the LMAT Grand database, at least a "fat node" will be needed with at least half TB of main memory and a SSD for storing the database.. Fortunately, other LMAT databases have more modest requirements. LMAT uses PERM, a 'C' library for persistent heap management developed at LLNL too. The PERM memory allocator replaces the standard 'C' dynamic memory allocation functions with compatible versions that provide persistent memory to application programs. For PERM (so LMAT) to work in the right conditions, some kernel tuning is advisable (administrative privileges required), which consists in turning off periodic flush to file and dirty ratio flush and switching off address space randomization:
```
> sudo sh -c 'echo 0 > /proc/sys/vm/dirty_
writeback_centisecs'
> sudo sh -c 'echo 100 > /proc/sys/vm/
dirty_background_ratio'
> sudo sh -c 'echo 100 > /proc/sys/vm/
dirty_ratio'
> sudo sh -c 'echo 0 > /proc/sys/kernel/
randomize_va_space'
```

16. LMAT can be built with different compilers and with different building profiles. The redoall script accepts two different optional parameters. The first one chooses the build profile of CMake: "D" for debug, "R" for release (this is the default), "I" for release with debugging info, and "M" for minimum executable size release. The second parameter selects the compiler family to use: "gnu" for using GNU gcc compiler, "intel" for Intel compilers, "clang" for clang/LLVM compiler with OpenMP support, and "ibmpwr9" for compiling in an IBM supercomputer with Power9 processors using the IBM compilers optimized for this architecture.

17. The LMAT Grand database (lmat-4-14.20mer.db) requires 462 GB of disk space, so it takes some time to be downloaded and assembled by the get_db script. It is not mandatory but quite advisable to move this huge database to another filesystem, especially not the same that will suffer the I/O stress

related to the data being processed by LMAT. Ideally, this filesystem will be allocated in a storage device with a very broad reading bandwidth and very low latency, like a SSD (solid-state drive). In this protocol, we suppose such a filesystem is mounted under /fastdisk.

18. The LMAT taxonomic classifier core (read_label) is a C++ program that requires several parameters. LMAT provides with a convenient launcher (run_rl.sh) that not only simplifies the call to read_label, but also launches other tools (tolineage and fsreport) to get some additional results. The pyLMAT_rl.py script in the ratsvm repository eases even more the process by automating the call to run_rl.sh for several related samples, which is our case.

19. The LMAT content summarization core (content_summ) is a C++ program requiring several parameters, so LMAT provides with a convenient launcher (run_cs.sh), which simplifies the call to content_summ and does some additional work. The pyLMAT_cs.py script in the ratsvm repository goes one step beyond to facilitate even more the process by automating the call to run_cs.sh for several related samples, like the ours.

20. The scripts rawlmat2lmat.py and lmat2cmplx.py perform a two-step translation of the LMAT results to the input format of cmplxcruncher. The goal of the former is to centralize the LMAT output in a single directory, in this case, the virome one. The mission of the latter is parsing all the files in the directory to extract the right taxonomic information for every time sample and write it in a spreadsheet in the format expected by cmplxcruncher. Both scripts have different parameters, whose default values are the convenient ones for this protocol, but can be changed if the protocol is to be used for different data.

21. complexCruncher computing kernel is a Python tool to analyze the dynamics of ranking processes in metagenomics. Cmplxcruncher requires Python 3.5 or later is needed and following python packages: pandas >1.16.0 with openpyxl, numpy, scipy, matplotlib >1.5.0 and xlrd >0.9. The tool will let you play with the data and repeat the analysis. You can see the graphs interactively or save them to disk in different formats. Alternatively, you can run the code in automatic (non-interactive mode) with the flag --automatic, like in the protocol: it tries to process every text or Excel file it could find in its data file tree. The results will be saved in a subdirectory of the data one, typically "results," which some subdirectories with self-descriptive names: fits, hist, and corrank, for correlation and rank plots. You can choose different names for the data and results directories, and select different levels and depth of analysis. Finally, be aware that complexcruncher is still under development.

References

1. Miller RR, Montoya V, Gardy JL et al (2013) Metagenomics for pathogen detection in public health. Genome Med 5(9):81
2. Ercolini D (2013) High-throughput sequencing and metagenomics: moving forward in the culture-independent analysis of food microbial ecology. Appl Environ Microbiol 79(10):3148–3155
3. Martí JM, Martínez-Martínez D, Rubio T et al (2017) Health and disease imprinted in the time variability of the human microbiome. mSystems 2(2):e00144-16
4. Wu H, Tremaroli V, Bäckhed F (2015) Linking microbiota to human diseases: a systems biology perspective. Trends Endocrinol Metab 26(12):758–770
5. Noecker C, Eng A, Srinivasan S et al (2016) Metabolic model-based integration of microbiome taxonomic and metabolomic profiles elucidates mechanistic links between ecological and metabolic variation. MSystems 1(1):13
6. Greenblum S, Turnbaugh PJ, Borenstein E (2012) Metagenomic systems biology of the human gut microbiome reveals topological shifts associated with obesity and inflammatory bowel disease. Proc Natl Acad Sci 109(2):594–599
7. Bashan A, Gibson TE, Friedman J et al (2016) Universality of human microbial dynamics. Nature 534(7606):259
8. Smith HF (1938) An empirical law describing heterogeneity in the yields of agricultural crops. J Agric Sci 28(1):1–23
9. Taylor LR (1961) Aggregation, variance and the mean. Nature 189(4766):732–735
10. Mantegna RN, Stanley HE (1995) Scaling behaviour in the dynamics of an economic index. Nature 376(6535):46
11. Eisler Z, Kertesz J, Yook S et al (2005) Multiscaling and non-universality in fluctuations of driven complex systems. Europhys Lett 69(4):664
12. De Menezes MA, Barabási A (2004) Fluctuations in network dynamics. Phys Rev Lett 92(2):028701
13. Reed DH, Hobbs GR (2004) The relationship between population size and temporal variability in population size. In: Anonymous animal conservation forum, vol 7. Cambridge University Press, p 1
14. Anderson RM, Gordon DM, Crawley MJ et al (1982) Variability in the abundance of animal and plant species. Nature 296(5854):245–248
15. Cohen JE, Xu M, Schuster WS (2013) Stochastic multiplicative population growth predicts and interprets Taylor's power law of fluctuation scaling. Proc R Soc Lond B Biol Sci 280(1757):20122955
16. Živković J, Tadić B, Wick N et al (2006) Statistical indicators of collective behavior and functional clusters in gene networks of yeast. Eur Phys J 50(1–2):255–258
17. Kendal WS (2003) An exponential dispersion model for the distribution of human single nucleotide polymorphisms. Mol Biol Evol 20(4):579–590
18. Zhang Z, Geng J, Tang X et al (2014) Spatial heterogeneity and co-occurrence patterns of human mucosal-associated intestinal microbiota. ISME J 8(4):881
19. Kaltz O, Escobar-Páramo P, Hochberg ME et al (2012) Bacterial microcosms obey Taylor's law: effects of abiotic and biotic stress and genetics on mean and variance of population density. Ecol Process 1(1):5
20. Ramsayer J, Fellous S, Cohen JE et al (2012) Taylor's law holds in experimental bacterial populations but competition does not influence the slope. Biol Lett 8(2):316–319
21. Pérez-Cobas AE, Artacho A, Ott SJ et al (2014) Structural and functional changes in the gut microbiota associated to Clostridium difficile infection. Front Microbiol 5:335
22. Ding T, Schloss PD (2014) Dynamics and associations of microbial community types across the human body. Nature 509(7500):357
23. Gajer P, Brotman RM, Bai G et al (2012) Temporal dynamics of the human vaginal microbiota. Sci Transl Med 4(132):132ra52
24. Nayfach S, Pollard KS (2016) Toward accurate and quantitative comparative metagenomics. Cell 166(5):1103–1116. https://doi.org/10.1016/j.cell.2016.08.007
25. Lusk RW (2014) Diverse and widespread contamination evident in the unmapped depths of high throughput sequencing data. PLoS One 9(10):e110808
26. Weiss S, Amir A, Hyde ER et al (2014) Tracking down the sources of experimental contamination in microbiome studies. Genome Biol 15(12):564. https://doi.org/10.1186/s13059-014-0564-2
27. Gruber K (2015) Here, there, and everywhere. EMBO reports:e201540822
28. Kircher M, Sawyer S, Meyer M (2012) Double indexing overcomes inaccuracies in multiplex sequencing on the Illumina platform. Nucleic Acids Res 40(1):e3
29. Thoendel M, Jeraldo P, Greenwood-Quaintance KE et al (2017) Impact of

contaminating DNA in whole-genome amplification kits used for metagenomic shotgun sequencing for infection diagnosis. J Clin Microbiol 55(6):1789–1801

30. Ames SK, Hysom DA, Gardner SN et al (2013) Scalable metagenomic taxonomy classification using a reference genome database. Bioinformatics 29(18):2253–2260
31. Ames SK, Gardner SN, Martí JM et al (2015) Using populations of human and microbial genomes for organism detection in metagenomes. Genome Res 25(7):1056–1067
32. Langmead B, Trapnell C, Pop M et al (2009) Ultrafast and memory-efficient alignment of short DNA sequences to the human genome. Genome Biol 10(3):R25
33. Anonymous (2016) Database resources of the National Center for biotechnology information. Nucleic Acids Res 44(D1):D19
34. Burguet-Castell J, Martí JM (2018) Entrez: A simple Python interface to the NCBI databases (https://github.com/jordibc/entrez)
35. Li H, Handsaker B, Wysoker A et al (2009) The sequence alignment/map format and SAMtools. Bioinformatics 25(16):2078–2079
36. Torvalds L, Hamano J (2010) Git: Fast version control system.
37. Minot S, Bryson A, Chehoud C et al (2013) Rapid evolution of the human gut virome. Proc Natl Acad Sci U S A 110(30):12450–12455

Chapter 18

Bioinformatics Assembling and Assessment of Novel Coxsackievirus B1 Genome

Jake Lin, Bryn Y. Kimura, Sami Oikarinen, and Matti Nykter

Abstract

The human microbiome project via application of metagenomic next-generation sequencing techniques has found surprising large and diverse amounts of microbial sequences across different body sites. There is a wave of investigators studying autoimmune related diseases designing from birth case and control studies to elucidate microbial associations and potential direct triggers. Sequencing analysis, considered big data as it typically includes millions of reads, is challenging but particularly demanding and complex is virome profiling due to its lack of pan-viral genomic signature. Impressively thousands of virus complete genomes have been deposited and these high-quality references are core components of virus profiling pipelines and databases. Still it is commonly known that most viral sequences do not map to known viruses. Moreover human viruses, particularly RNA groups, are notoriously heterogeneous due to high mutation rates. Here, we present the related assembling challenges and a series of bioinformatics steps that were applied in the construction of the complete consensus genome of a novel clinical isolate of Coxsackievirus B1. We further demonstrate our effort in calling mutations between prototype Coxsackievirus B1 sequence from GenBank and serial clinical isolate genome grown in cell culture.

Key words Genomics, Assembly, Bioinformatics, Enterovirus, Coxsackievirus, T1D

1 Introduction

Coxsackievirus B (CVB), with six serotypes and a member of the human enterovirus genus, have long interested researchers due to its pathogenic role in gastrointestinal inflammation and implications with cardiomyopathy. Coxsackievirus B have also been implicated in Type 1 diabetes (T1D), which is an autoimmune disorder onset typically in young children. The trigger is unknown, but clinical T1D occurs when about 90% pancreatic islet beta cell population is destroyed and thereafter regular external insulin is required to regulate blood glucose levels to avoid hyperglycaemia leading to organ damage and death. The detection of islet autoantibodies in the blood signals autoimmunity and multiple autoantibody confirmations accurately predict T1D. The window of time between autoimmunity and T1D onset can vary from months to multiple years [1]. Two of the

Andrés Moya and Vicente Pérez Brocal (eds.), *The Human Virome: Methods and Protocols*, Methods in Molecular Biology, vol. 1838, https://doi.org/10.1007/978-1-4939-8682-8_18, © Springer Science+Business Media, LLC, part of Springer Nature 2018

autoantibodies namely insulin (IAA) and glutamic acid decarboxylase (GADA) may predict different progression models from the triggering the disease to onset of T1D. It has been reported that IAA, peaking in the first 3 years of child's life, is associated with early-age T1D onset and GADA confirmation incidence plateaus into adolescence, suggesting divergent trends and possibly multiple environmental triggers and immune interactions. Enterovirus sequences have been found in pancreas tissue [2] donated from recent T1D patients. CVB with six serotypes is the subgroup of enterovirus most cited in PubMed associated with T1D. While recent advances in metagenomic shotgun sequencing together with lowered cost of sequencing have allowed investigators to sequence virus directly from the environment, for example stool or blood samples, virus profiling and particularly genotype identification are challenging due to lack of universal viral marker, insufficient amount of preference genomes, and also large genetic distance from preference virus sequences. The genetic distance is exasperated by their innate high pace of mutation, particularly in RNA viruses including CVB. This report presents a set of bioinformatics procedures, known and novel, toward building a novel complete CVB1 consensus genome obtained from clinical virus isolate grown in pancreatic cell culture model using next-generation sequencing and multiple time points.

2 Materials

A novel wild Coxsackievirus B1 clinical isolate strain was continuously cultured in human pancreas ductal (Panc-1, abbreviated as A) and pancreas islet (1.1B4, abbreviated as B) cells to establish a persistent infection model for CVB1. We note that consensus genome assembly and variant calling are applied for Panc-1 (A) samples where quality control and contig generation were done for both cell types. The infected cells were grown in +37 °C incubator with 5% CO_2 and treated three times a week by washing and adding fresh cell culture media [3]. The cells were monitored by microscopy and culture supernatant medium was harvested once a week and stored at −80 °C for sequencing analysis.

Three time points (a prep after 1 day of culture (time point 0), and then approximately 6 (time point 1) and 12 months (time point 2) after initiation of the virus culture) of harvested and frozen cell culture mediums containing virus were selected for NGS analysis. The virus was enriched in the cell culture and therefore, the virus RNA was extracted directly for cell culture supernatant for next-generation sequencing analysis. The sample preparation was done according to the protocol presented also in this book [Kramna and Cinek, "Virome sequencing of stool samples"]:

Selected known references.

CVB1 Prototype (GenBank accession no NC_001472.1).

CVB1 wild type (GenBank accession no AY186745.1).

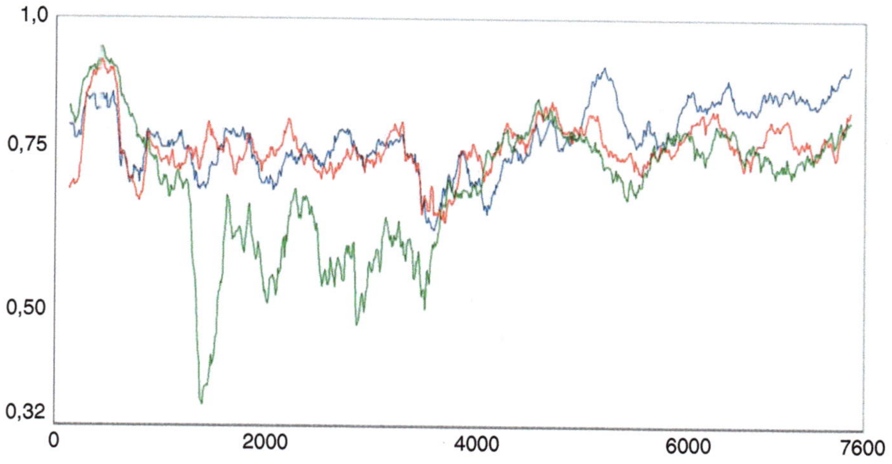

Fig. 1 Reference similarity plots. Set against CVB1 prototype, CVB1 clinical isolate in red and CVB1 wild type in blue are compared. CVB5 in green confirms the large genetic diversity in structural capsid regions (~800–3000) between serotypes

Table 1
Bioinformatics dependencies (tested in Linux environment) and applied tools

Name	Purpose
Bedtools [4]	Tool for genome arithmetic
BWA [5]	Short read alignment
FastQC [6]	Read quality report util
IGV [7]	Aligned track visualization
KmerGenie [8]	Sequence utility
Samtools [9]	Sequence utility
Taxonomer [10]	Virome profiler, no assembly, direct mapping
Velvet [11]	Short read assembler
Vipie [12]	Multi-sample virome profiler; de novo assembly

CVB1 Clinical isolate (Hyöty lab, University of Tampere, unpublished).

CVB5 Prototype (GenBank accession no MF973166.1).

Figure 1 demonstrates the substantial sequence differences between CVB1 prototype and CVB1 wild type, exposing the need for an updated and specific clinical isolate. Variant analysis between virus sequences analyzed from different time points in cell culture also mandates relevant consensus reference sequences (Table 1).

Custom code requires Python (Version 3.4+) and pysam [4, 13].

Detailed directions and README file are accessible at: https://sourceforge.net/projects/contig-mutation-caller

3 Methods

3.1 Virome Profiling and Assessment

Preliminary quality control (QC) and virus profiling were done in Vipie [12], a web-based virome NGS pipeline. Resultant virome profiles from CVB1 prototype samples (Fig. 2a) show that a clear majority, over 99% of viral reads mapped to its homologous sequence in the GenBank (accession NC_001472.1), while for clinical virus isolate (Fig. 2b) grown in pancreatic ductal and islet cells, the mapped profile accessions are highly diverse. The profile includes more than 20 different matched EV accessions with proportions

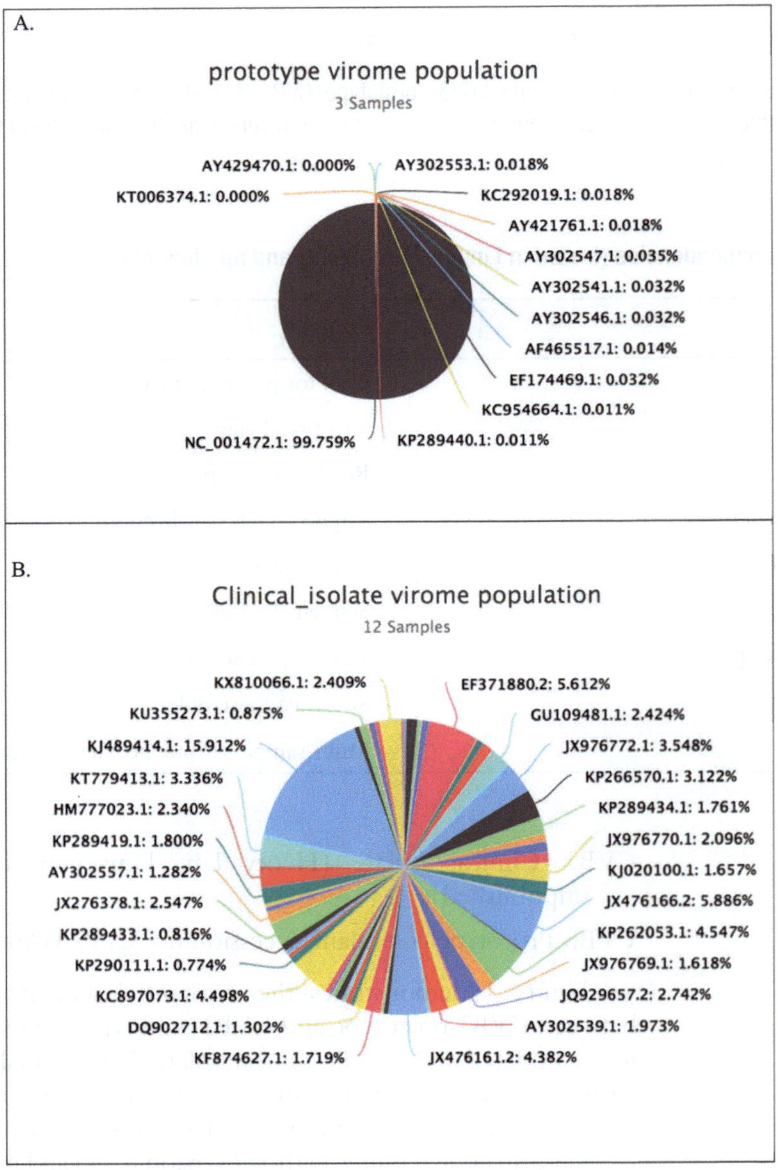

Fig. 2 Vipie virome population profiling—known CVB1 prototype strain (**a**) and wild CVB1 clinical isolate (**b**)

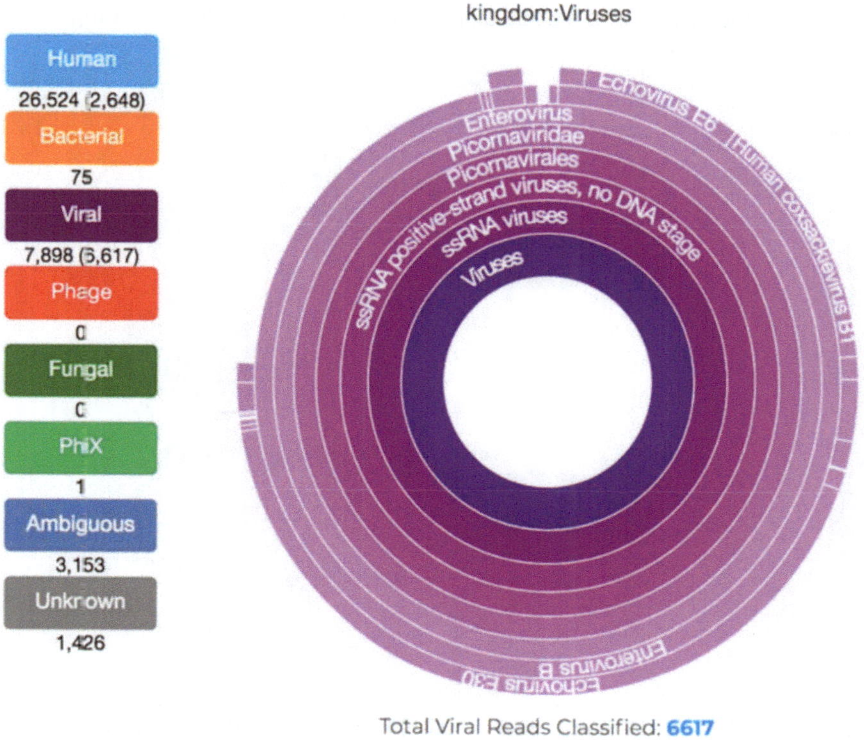

Fig. 3 Virome profile validation of clinical isolate of CVB1-Taxonomer pipeline operates on one sample at a time and confirms CVB1 but also other EV accessions

between 1% and 15%. The heterogeneous and divergent profiles are expected in the analysis of clinical virus isolate and confirmed the need for enhanced custom methods and also motivate the need for novel genome for accurate and sensible mutation assessments.

Vipie profiled results were confirmed using virome profiling pipeline Taxonomer. Though Taxonomer restricts submissions to one sample (two paired end sequences files), the accurate and easy to use tool produces attractive and intuitive results. Clinical CVB1 isolate sequence from time point 0 was analyzed using Taxonomer, the results reported CVB1 along with two Echovirus strains (Fig. 3). The amount of human, bacterial, and unknown viral reads identified are consistent with Vipie aligned distribution report. Taxonomer tool is ideal for fast metagenomics profiling as it maps the inputs reads directly onto known references but as the tool skips de novo assembly, it is not suitable for consensus construction as contigs are not produced.

3.2 QC and De Novo Assembly

Analysis of changes occurring in the CVB1 virus genome over time during infection of pancreas model cells was carried out to investigate the possible role of CVB1 in Type 1 diabetes. To compare changes between time points a consensus sequence of the genome for each time point was created from fastq files generated through

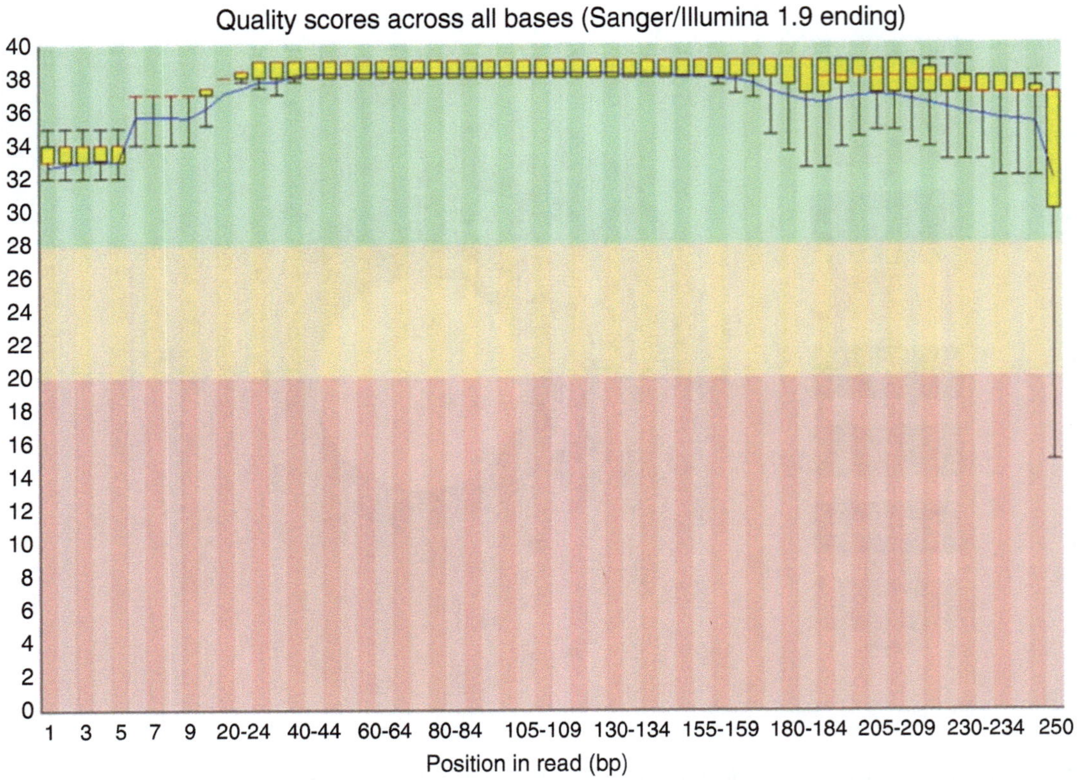

Fig. 4 A per base quality score plot from FastQC. The x-axis indicates insert positions along the reads and the y-axis represents possible quality score ranges. A box-and-whisker plot of quality scores is calculated for each position or range of positions across all reads. The median quality score stays well above 28, the cutoff for "very good quality scores" until the very end of the insert length

NGS. This section will discuss how the clinical isolate samples were processed from quality control of the fastq files to reporting changes across time points. The fastq files from clinical sourced CVB1 were examined in FastQC and determined to have acceptable quality so no trimming was done on the reads (Fig. 4). To assess quality of the sequenced reads, the phred quality scores were examined for each position in the reads. The phred score is −10 log10 Probability {nucleotide was correctly identified}. Therefore, a phred score of 20 means there is a 99% probability the nucleotide was correctly identified during sequencing.

An important component of virome sequencing analysis involves estimation of potential bacterial and human mapped reads. Figure 5 shows that there are partial reads mapping to human and bacterial 16S ribosome marker, a majority of sample reads are classified as dark viral matter. The high number of dark viral proportions pinpoint the lack of sufficient references and also greater genetic distance between clinical strains and references due to higher evolutionary mutation rates [14].

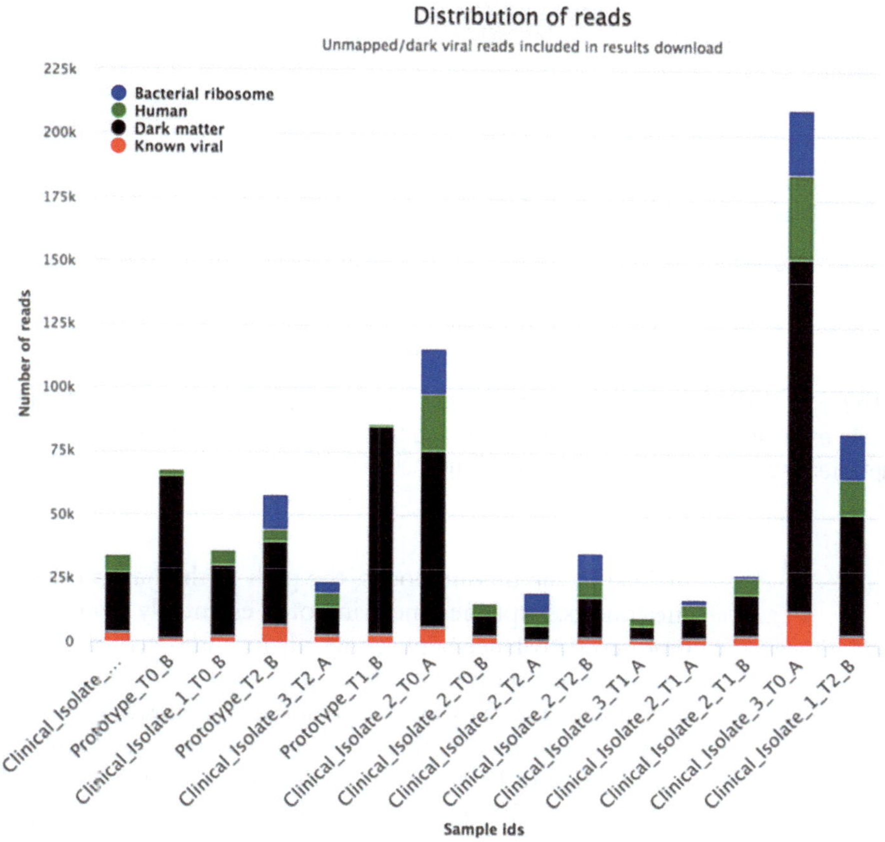

Fig. 5 Mapped distribution report produced from Vipie reports partial virome reads mapping to bacterial and human references. Because of lack of updated genome references and high mutation rates, it can be seen that a majority of reads, in black, are classified as unknown viral reads (Dark matter)

Sanger sequencing was done on the clinical isolate particles when they first arrived at the research facility and a nearly complete genome sequence was produced. This sequence was used as the reference to help assemble a consensus sequence from for the first time point. The steps to assemble a consensus sequence involved first creating contigs (a sequence created from overlapping reads) from the fastq reads, aligning the contigs to a reference, and using the nucleotide sequences from the contigs to form the consensus. Vipie uses Velvet by default, a de Bruijn graph-based [15] de novo assembler, to assemble contigs from short-read fastq files produced by NGS sequencing. The program KmerGenie was used to estimate the optimal k-mer size and average coverage cut-off parameters used during de novo assembly. In addition, the minimum contig length and average read size for each sample from the fragment analyzer were given as parameters to the assembler. Velvet assembly consists of two steps. Operating on quality filtered and interlaced read files, the first step creates a "hash" set of k-mer sized length nodes, storing all (k-1) overlapped nucleotides. The second step attempts to walk the nodes

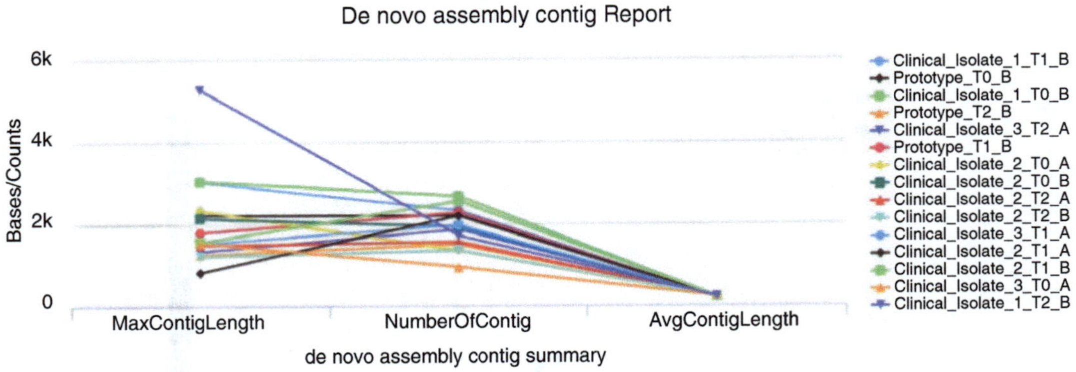

Fig. 6 Contig report. Contig outputs, with maximum size ranges around 2000–5500 and average length of 300 are shown. An estimated 2000 contigs are produced and based on Vipie QC read report (not shown). Contig sizes do not directly correlate with raw number of sequence reads

by linking the unique nodes, the path of the hash nodes represents the contigs. Explained more in notes, essentially k-mer size selection represents a balance between specificity and precision as larger k-mer sizes increase specificity due to increased overlaps and at the same time lower sensitivity as more reads are discarded. Generated CVB1 contig report from Vipie is shown in Fig. 6 where each sample produced roughly 1200–2500 contigs and a sample maximum length of more than 5000. Contig results are also stored as separate fasta files for each sample and accessible directly from Vipie secured results where they can be downloaded.

3.3 Consensus Construction

The contigs were aligned to the reference strain using BWA producing alignment files in the sam file format. The amount of the reference covered by each set of contigs was recorded. In addition, contigs with mapping quality below MAPQ score of 20 were removed. Consensus sequences were created for all virus strains collected at the three different time points of the experiment. The time point 0 sequence was used as a reference for time point 1 and time point 1 for time point 2. It is likely that this approach will work and be sustainable for studies with denser and higher number of serial samples. Time point $(i + 1)$ consensus sequences were made using the sequences from the contigs in the order that they were aligned to the reference at i. Vipie by default produces contigs using Velvet, the contigs are scaffolds, implying nonoverlapping. For each referential position covered by a contig, a nucleotide was appended to the consensus sequence. In scenarios when differences between the contig and the reference were found, the nucleotides from the contigs were appended to the targeted consensus sequence. The result was a sequence identical to the reference except for substitutions and indels found in the relevant time point contigs. The consensus contig coverages and assembly parameters generated from

Table 2
Contig coverage of reference

Strain time point	Kmer/coverage	Contig genome coverage relative to latest consensus
Clinical isolate 1 :0	21/2	96.4%
Clinical isolate 1 :1	25/2	91.7%
Clinical isolate 1 :2	31/2	80.2%

KmerGenie for all three time points are shown in Table 2 and the steps below detail the described approach:

3.3.1 De Novo Assembly Parameters

Make a kgen.in **file for paired-end reads that contains the names of the forward and reverse** fastq files:

```
vim kgen_8.in
```

KmerGenie to determine the correct hash length and cutoff:

```
/kmergenie-1.7044/kmergenie kgen_2.in
```

3.3.2 Serial Based Consensus Comparison and Construction

Step 1: Create bwa index from consensus sequence from previous point or reference sequence.

```
bwa index--p Clinical_Isolateref Clinical_
Isolateref.fa
```

Step 2: Align the contigs from the current point to the consensus sequence from previous time point or reference. Result is an alignment file describing where contigs aligned along the previous time point sequence.

```
bwa mem--t 3--L 6,6 /path/ref_Seq/
Clinical_Isolateref Timepoint_[i]_contigs.
fa > Timepoint_[i]_alignedto_Clinical_
Isolateref.sam
```

Step 3: Extract contigs with MAPQ > = 20 (adjustable via parameter).

```
/usr/bin/python2.6 /path/extract_sam_ac-
cession_simp.py Timepoint_[i]_aligned-
to_Clinical_Isolateref.sam Timepoint_[i]_
alignedto_Clinical_Isolateref_extracted.
sam 20
```

Visualization-- IGV

Step 4: Sort contigs according to start position.

```
igvtools sort Timepoint_[i]_aligned-
to_Clinical_Isolateref_extracted.
sam Timepoint_[i]_alignedto_Clinical_
Isolateref_extracted_sorted.sam
```

Step 5: Create an index of the contig alignment sam file.

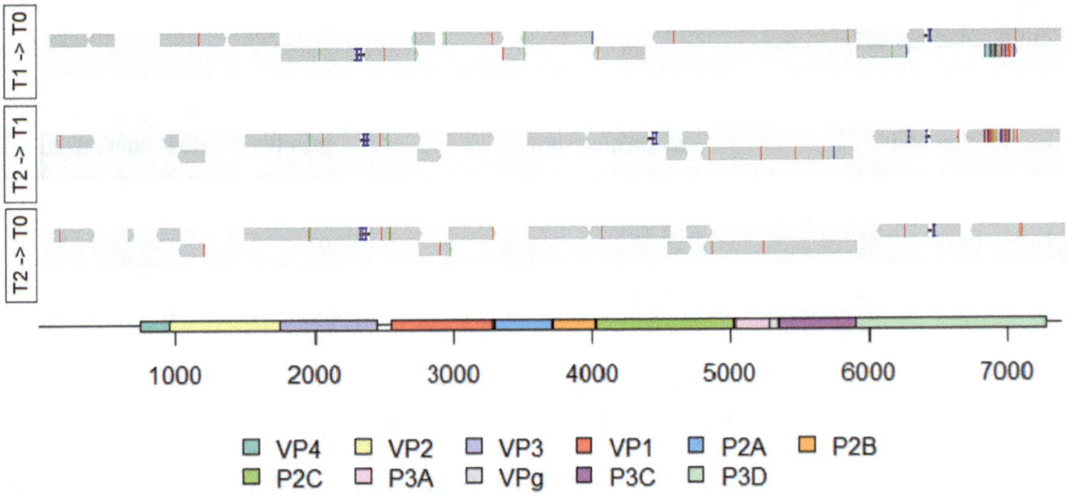

Fig. 7 Contigs from time point 0 are shown aligned to the clinical isolate reference genome in the top track. Time point 1 contigs are aligned to the time point 0 consensus sequence in the middle track and time point 2 contigs are aligned to the time point 1 consensus sequence in the bottom track

```
igvtools index Timepoint_[i]_alignedto_
Clinical_Isolateref_extracted_sorted.sam
```

Step 6: Determine differences between timepoints and construct a consensus sequence for the current time point.

```
python /mutation_caller_V4.py Timepoint_
[i]_alignedto_Clinical_Isolateref_ex-
tracted_sorted.sam /path/ref_Seq/
Clinical_Isolateref.fa /path/Contigs/clini-
cal_isolate/mut_seqs_clinical_isolate
```

Step 7: Repeat with next time point(Fig. 7).

3.4 Variant Calling

Comparisons were made between the prototype sequence and different time points of clinical isolate. A script was written in python that tabulates single nucleotide polymorphisms (SNPs) and indels (insertions and deletions). The script made use of pysam, a module that creates easy access to nucleotide read information from an aligned (.sam) file. During the comparison between the reference and time point 0, each contig from the time point 0 sample was examined for indels from nucleotide differences compared to the window region of the reference to which it aligned. These differences, relatively consistent between time points, are reported in Table 3.

It has been reported that CVB3 variants and deletions within 5′ terminal [16–18] impact virulence and are also of great interest pertaining to this project. The variant results thus far have not matched, and probably should not since this study is using novel clinical isolate CVB1, a different serotype but warranting considerably more analysis. The fact also is that the variants reported are based on majority consensus but minor allele fractions, while less

Table 3
The total number of SNPs and indels found in clinical isolate serial samples grown in PANC-1 cells

	Clinical Isolate 1T0	Clinical Isolate 1T1	Clinical Isolate 1T2
SNPs	15	24	18
Deletions	1	0	3
Insertions	0	1	2

study, have important biological roles including associated cancer odds ratios [19]. The tools and steps introduced are largely applicable for minor allele study and benefits from the constructed consensus. These steps will also benefit from additional analysis to include comprehensive translation and classification of variant functional and potential important untranslated region (UTR) impacts.

4 Notes

1. CVB1 was used in this study because of possible association to T1D. It also represents a good model for novel virus assembly. CVB5 was selected as an example strain to demonstrate diverse EV sub-serotype genetic distances.
2. Pancreatic ductal and islet cell lines were selected as they are known to be able to establish a persistent infection model for CVB1 and highly relevant for T1D.
3. Virome sample profiling was validated with VirusTAP [20]. The tool is web based and includes de novo assembly and outputs contigs. As VirusTap removes all human and bacterial reads prior to assembly, it is not fit for consensus.
4. In the scenario within consensus construction where particular samples exhibit magnitudes more contamination or dark matter reads relative to other samples, the contigs generated likely will in effect possess similar proportions. Remapping directly to virus genomes of interest with a low MAPQ score cutoff and then using the aligned reads for de novo assembly can assist with building more desirable contigs.
5. KmerGenie [8] is recommended for optimizing k-mer size, "k," and k-mer coverage cutoff for Velvet and all de Bruijn graph solutions. K-mer size must be odd and smaller than read insert length. The program estimates the k-mer size that maximizes the number of unique k-mers found in the reads in a fastq file or set of paired fastq files produced from NGS sequencing. In a de Bruijn graph, these k-mers can be linked by finding common prefixes and suffixes thus forming contiguous regions of the original genome. Shorter k-mers produce

more connections between reads increasing the proportion of correctly connected reads (high sensitivity) but also increasing the proportion of incorrectly connected reads (low specificity). In contrast, longer k-mers produce fewer connections decreasing the proportion of correctly connected reads but increasing the proportion of correctly unconnected reads. The k-mer coverage is the number of times the k-mer was observed in the reads. A k-mer coverage cutoff determines the number of times a k-mer must be seen in the reads before it is used in the assembly.

References

1. Pociot F, Lernmark Å (2016) Genetic risk factors for type 1 diabetes. Lancet 387(10035):2331–2339
2. Krogvold L, Edwin B, Buanes T et al (2015) Detection of a low-grade enteroviral infection in the islets of langerhans of living patients newly diagnosed with type 1 diabetes. Diabetes 64(5):1682–1687
3. Sane F, Caloone D, Gmyr V et al (2013) Coxsackievirus B4 can infect human pancreas ductal cells and persist in ductal-like cell cultures which results in inhibition of Pdx1 expression and disturbed formation of islet-like cell aggregates. Cell Mol Life Sci 70(21):4169–4180
4. Quinlan AR, Hall IM (2010) BEDTools: a flexible suite of utilities for comparing genomic features. Bioinformatics 26(6):841–842 https://doi.org/10.1093/bioinformatics/btq033
5. Li H, Durbin R (2009) Fast and accurate short read alignment with burrows-wheeler transform. Bioinformatics 25:1754–1760
6. Babraham Bioinformatics, FastQC. https://www.bioinformatics.babraham.ac.uk/projects/fastqc/. Accessed 02 2018
7. Robinson JT, Thorvaldsdóttir H, Winckler W et al (2011) Integrative genomics viewer. Nat Biotechnol 29:24–26
8. Chikhi R, Medvedev P (2013) Informed and automated k-Mer size selection for genome assembly, HiTSeq
9. Li H, Handsaker B, Wysoker A et al (2009) 1000 genomes project, the sequence alignment/map format and SAM tools. Bioinformatics 25(16):2078–2079
10. Flygare S, Simmon K, Miller C et al (2016) Taxonomer: an interactive metagenomics analysis portal for universal pathogen detection and host mRNA expression profiling. Genome Biol 17(1):111
11. Zerbino DR, Birney E (2008) Velvet: algorithms for de novo short read assembly using de Bruijn graphs. Genome Res 18:821–829
12. Lin J, Kramna L, Autio R et al (2017) Vipie: web pipeline for parallel characterization of viral populations from multiple NGS samples. BMC Genomics 18(1):378
13. pysam. https://github.com/pysam-developers/pysam. Accessed 02 2018
14. de Bruijn NG (1946) A combinatorial problem. Koninklijke Nederlandse Akademie v Wetenschappen 49:758–764
15. Attar N (2016) Viral evolution: more of the world's a phage. Nat Rev Microbiol 14:269
16. Massilamany C, Gangaplara A, Basavalingappa RH et al (2015) Mutations in the 5' NTR and the non-structural protein 3A of the Coxsackievirus B3 selectively attenuate myocarditogenicity. PLoS One 10(6):e0131052
17. Chapman NM, Kim KS, Drescher KM et al (2008) 5′ terminal deletions in the genome of a coxsackievirus B2 strain occurred naturally in human heart. Virology 375(2):480–491
18. Rinehart JE, Gómez RM, Roos RP (1997) Molecular determinants for virulence in coxsackievirus B1 infection. J Virol 71(5):3986–3991
19. Pomerantz MM, Freedman ML (2011) The genetics of cancer risk. Cancer J 17(6):416–422
20. Yamashita A, Sekizuka T, Kuroda M (2016) VirusTAP: viral genome-targeted assembly pipeline. Front Microbiol 7:32

Index

Andrés Moya and Vicente Pérez Brocal (eds.), *The Human Virome: Methods and Protocols*, Methods in Molecular Biology, vol. 1838, https://doi.org/10.1007/978-1-4939-8682-8, © Springer Science+Business Media, LLC, part of Springer Nature 2018

MIX
Papier aus verantwortungsvollen Quellen
Paper from responsible sources
FSC® C105338

FSC
www.fsc.org

If you have any concerns about our products,
you can contact us on
ProductSafety@springernature.com

In case Publisher is established outside the EU,
the EU authorized representative is:
Springer Nature Customer Service Center GmbH
Europaplatz 3, 69115 Heidelberg, Germany

Printed by Libri Plureos GmbH
in Hamburg, Germany